T0132442

Electrostatic Considerations in Mitosis

by

L. John Gagliardi

Rutgers University
Department of Physics
Camden, NJ 08102
email: gagliard@camden.rutgers.edu
web: http://crab.rutgers.edu/~gagliard

iUniverse books may be ordered through booksellers or by contacting:

iUniverse
1663 Liberty Drive
Bloomington, IN 47403
www.iuniverse.com
1-800-Authors (1-800-288-4677)

Because of the dynamic nature of the Internet, any Web addresses or links contained in this book may have changed since publication and may no longer be valid. The views expressed in this work are solely those of the author and do not necessarily reflect the views of the publisher, and the publisher hereby disclaims any responsibility for them.

ISBN: 978-1-4401-7378-3 (sc)
ISBN: 978-1-4401-7376-9 (hc)
ISBN: 978-1-4401-7377-6 (ebook)

Printed in the United States of America

iUniverse rev. date: 10/22/2009

I would like to thank my wife Grace for her encouragement and considerable help in writing this book.

Preface

About ten years ago it became clear to me that endogenous intracellular electric fields can be extended over cellular distances by microtubules. The implications of this are that electrostatic force, one of the most powerful forces in nature, can operate in cells in spite of counterion screening. This realization, coupled with an evolutionary perspective regarding mitosis, led to the development of an electrostatic force producing mechanism for mitotic motions. This, in turn, suggested that attempting a minimal assumptions model for post-attachment chromosome motions might be worth the effort.

My focus within this approach has been on known, as well as assumed, cellular charge distributions. The known charge is at the ends of microtubules and on chromosome arms; assumed charge distributions have included positive charge on kinetochores and negative charge on centrosome matrices. Given the electrostatic interactions associated with these distributions, a minimal assumptions model of post-attachment prometaphase, metaphase, and anaphase chromosome motions was envisioned.

It has been implicit in a series of my publications in Physical Review E and the Journal of Electrostatics exploring the physics of these motions that a minimal assumptions model of this sort, if internally consistent, could lead to the eventual discovery of the specific molecules in kinetochores and centrosome matrices that are responsible for mitotic chromosome motions. This is what is happening. Currently, electrostatic force, due to positive charge at kinetochores, has been implicated in kinetochore-microtubule dynamic coupling during mitosis. In addition, recent experiments have revealed that centrosomes are negatively charged.

This book proposes a comprehensive model for post-attachment chromosome motions based on microtubule dynamics, the electrical properties of (tubulin and) microtubules, and the charge on chromosome arms, kinetochores and centrosomes. Although such approaches are not the norm in biology, it can be argued that the time may have come for "discontinuous leaps" involving comprehensive theoretical models to play a larger role in cell biology. Indeed, the history of science teaches that many (if not most) major scientific breakthroughs have come in this manner.

Since the physics in this book has been refereed in a series of publications, it is my hope that cell biologists can rest assured that they need not worry so much about the physics and can therefore feel free to concentrate on the biology.

Contents

Chapter 1

Introduction

1.1 Preliminary considerations

Although the current paradigm in cell biology has been quite successful in solving a number of problems in the understanding of cell structure and function, important problems remain open. Some of the most notable of these are in the area of cell division. Given the constraints wittingly or unwittingly imposed by the molecular biology paradigm, it would seem that models of mitotic motions and events have become more and more complex, and therefore – to this observer at least – more and more unsatisfactory.

It may be helpful at this point to give an example. It is a common parlor trick to attract small bits of paper by a charged hard rubber hair comb. This motion is simply explained in terms of the known negative charge on the comb attracting the nearer induced positive charge on the bits of paper with the paper bits remaining overall electrically neutral and the displaced negative charges farther from the comb. There is no need to identify the specific molecules in the paper (or in the comb, for that matter) that are responsible for the attractive force and the subsequent motion. In fact, since the charged comb will attract many different kinds of paper as well as many other substances, attempting to identify the molecules in the various substances is counterproductive to explaining what is happening. Similarly, as will be discussed later, chromosome and other mitotic motions can be described in terms of electric charge distributions. However before this can be done, one must first establish that electrostatic force is significant within cells in spite of counterion screening.

The electromagnetic interaction is primarily responsible for the structure of matter from atoms to objects. Much of physics, all of chemistry, and most of biology are in this realm of sizes. Primitive eukaryotic cells had to divide prior to the evolution of very many biological mechanisms, and it is reasonable to assume that basic physics and chemistry played dominant roles in both mitosis (nuclear division) and cytokinesis (cytoplasmic division). It is proposed that nanoscale electrostatic interactions played a major role in the dynamics of cell division in primitive cells, and that the fundamental solutions to the problem of cell division that were found by primitive cells may largely persist in modern eukaryotic cells.

The mitotic spindle is responsible for the segregation of sister chromatids during cell division. Chromosomes are attached to the spindle with their kinetochores [Euteneuer and McIntosh, 1981] attached to the *plus* ends of microtubules [Rieder, 1982; Bergen *et al.*, 1980]. Chromosome movement is dependent on kinetochore-microtubule dynamics: a chromosome can move toward a pole only when its kinetochore is connected to microtubules emanating from that pole [Nicklas and Kubai, 1985]. A number of experimental studies have been undertaken to obtain information regarding microtubule dynamics, force production, and kinetochore function in mitotic cells. These experiments have revealed that the spindle can produce more force than is actually required to move a chromosome at the observed speeds for post-attachment movements, and that the force for the poleward motion of chromosomes can be localized at or near kinetochores [Nicklas, 1983; Mitchison *et al.*, 1986; Gorbsky *et al.*, 1987; Nicklas, 1989; Mitchison, 1989; Alexander and Rieder, 1991; Inoue and Salmon, 1995] or at spindle poles [Mitchison and Salmon, 1992; Maddox *et al.*, 2002; Zhang and Chen, 2003]. Quite some time ago, M. S. Cooper addressed a possible link between endogenous electrostatic fields and the eukaryotic cell cycle [Cooper, 1979]. An early review by Jaffe and Nuccitelli [Jaffe and Nuccitelli, 1977] focused on the possible influence of relatively steady electric fields on the control of growth and development in cells and tissues.

Microtubules continually assemble and disassemble, so the turnover of tubulin is ongoing. The characteristics of microtubule lengthening (polymerization) and shortening (depolymerization) follow a pattern known as "dynamic instability": that is, at any given instant some of the microtubules are growing, while others are undergoing rapid breakdown. In general, the rate at which microtubules undergo net assembly – or disassembly – varies with mitotic stage [Alberts *et al.*, 1994a].

Changes in microtubule dynamics are integral to changes in the motions of chromosomes during the stages of mitosis. Poleward and antipoleward chromosome motions occur intermittently during prometaphase and metaphase. Antipoleward motions dominate during the *congressional* movement of chromosomes to the cell *equator*, and poleward motion prevails during anaphase-A.

Chromosome motion during anaphase has two major components, designated as anaphase-A and anaphase-B. The poleward movement of anaphase-A is accompanied by the shortening of kinetochore microtubules at kinetochores and/or spindle poles. The second component, referred to as anaphase-B, involves the separation of the poles. Both components contribute to the increased separation of chromosomes during mitosis. It is proposed in this book that these changes in chromosome motions during mitosis can be attributed to changes in microtubule dynamics based on electrostatics. It is further proposed that the influence of intracellular pH changes on kinetochore microtubule dynamics is primarily responsible for post-attachment prometaphase and metaphase chromosome motions.

9

1.2 Some cellular electrostatics

In the cytoplasmic medium (cytosol) within biological cells, it has been generally thought that electrostatic fields are subject to strong attenuation by screening with oppositely charged ions (counterion screening), decreasing exponentially to much smaller values over a distance of several *Debye lengths*. The Debye length within cells is typically given to be of order 1 nm [Benedek and Villars, 2000a], and since cells of interest in the present work (i.e. eukaryotic) can be taken to have much larger dimensions, one would be tempted to conclude that electrostatic force could not be a major factor in providing the cause for mitotic chromosome movements in biological cells. However, the presence of microtubules, as well as other factors to be discussed shortly, change the picture completely.

Microtubules can be thought of as intermediaries that extend the reach of the electrostatic interaction over cellular distances, making this second most potent force in the universe available to cells in spite of their ionic nature. A presentation of some background material in electrostatics is summarized in Appendix A and a brief introduction to Debye lengths and counterion screening is given in Appendix B.

Microtubules are 25 nm diameter cylindrical structures comprised of *protofilaments*, each consisting of tubulin dimer subunits, 8 nm in length, aligned lengthwise parallel to the microtubule axis. The protofilaments are bound laterally to form a sheet that closes to form a cylindrical microtubule. The structure of microtubules is similar in all eukaryotic cells. Cross sections reveal that the wall of a microtubule consists of a circle of 4 to 5 nm diameter subunits. In most cases, the circle contains 13 subunits; however, 11, 12, 14, or 16 have also been observed. Neighboring dimers along protofilaments exhibit a small (B-lattice) offset of 0.92 nm from protofilament to protofilament, as depicted in Figure 1.1.

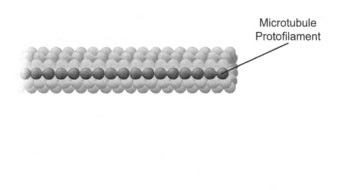

Fig. 1.1. A microtubule highlighting a protofilament. A B-lattice microtubule.

This offset will be approximated as 1 nm in the calculations in subsequent chapters since protofilament curling distributions for disassembling microtubules are more significant in determining the distances of protofilament free ends from various cellular structures. Protofilament curling of a disassembling microtubule is depicted in Figure 1.2.

Experiments have shown that the intracellular pH (pH_i) of many cells rises to a maximum at the onset of mitosis, subsequently falling during the later stages [see, for example, Steinhardt and Morisawa, 1982; Amirand *et al.*, 2000].

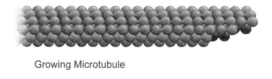

Growing Microtubule

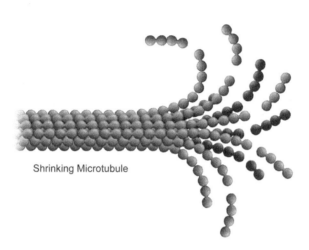

Shrinking Microtubule

Fig. 1.2. Shrinking (showing protofilament curling) and growing microtubules.

Although it is experimentally difficult to resolve the exact starting time for the beginning of the decrease in pH_i during the cell cycle, it appears to decrease 0.3 to 0.5 pH units from the typical peak values of 7.3 to 7.5 measured earlier during prophase [Steinhardt and Morisawa, 1982].

Studies [Schatten *et al.*, 1985] have shown that *in vivo* microtubule growth (polymerization) is favored by higher pH values. It should be noted that *in vitro* studies of the role of pH in regulating microtubule assembly indicate a pH optimum for assembly in the range of 6.3 to 6.4. The disagreement between *in vitro* and *in vivo* studies has been analyzed in relation to the nucleation potential of microtubule organizing centers like centrosomes [Schatten *et al.*, 1985], and it has been suggested that pH_i regulates the nucleation potential of microtubule organizing centers [Kirschner, 1980; De Brabander *et al.*, 1982; Deery and Brinkley, 1983]. This favors the more complex physiology characteristic of *in vivo* studies to resolve this question. It will therefore be assumed in this book that *in vivo* experimental design is more appropriate for experiments relating to pH conditions affecting microtubule assembly.

A number of investigations have focused on the electrostatic properties of microtubule tubulin subunits [Satarić *et al.*, 1993; Brown and Tuszyński, 1997; Baker *et al.*, 2001; Tuszyński *et al.*, 1998]. Large scale calculations of the tubulin molecule have been carried out using molecular dynamics programs along with protein parameter sets. The dipole moment of tubulin has been calculated to be as large as 1800 Debye (D) [Brown and Tuszyński, 1997; Tuszyński *et al.*, 1995]. Experiments [Tuszyński *et al.*, 1995; Sackett, 1997] have shown that tubulin net charge depends strongly on pH, varying quite linearly from –12 to –28 (electron charges) between pH 5.5 and 8.0. This could be significant for microtubule dynamics during mitosis because, as noted above, many cell types exhibit a decrease of 0.3 to 0.5 pH units from a peak at prophase during mitosis.

It has been determined that tubulin has a large overall negative charge of 20 at pH 7, and that as much as 40 % of the charge resides on C-termini. The C-termini can extend perpendicularly outward from the microtubule axis as a function of pH_i. It would seem reasonable to assume that an increased tubulin charge and the resulting greater extension of C-termini may be integral to an increased probability for microtubule assembly during prophase when pH_i is highest. A higher pH_i during prophase is consistent with increased interaction between the highly extended C-termini of tubulin dimers with appropriate regions of other nearest neighbor dimers. Given a decrease in pH_i during mitosis, changes in microtubule assembly probabilities – in conjunction with nanoscale electrostatic interactions – could be responsible for the observed changes in chromosome motions during mitosis. In particular, a decrease in pH_i during mitosis may act as a master clock controlling microtubule disassembly to assembly (disassembly/assembly) probability ratios during the phases of mitosis, thereby controlling the timing and dynamics of mitotic chromosome movements through metaphase. This will be discussed in more detail in Chapter 5 after the necessary groundwork has been developed.

It is generally accepted that the charge on the plus free ends of microtubules proximal to kinetochores is negative. (According to existing convention, these ends are designated *plus* because of their more rapid growth, there being no reference to charge in the use of this nomenclature.) Because of the electric dipole nature of the tubulin dimer subunits comprising microtubules, the net charge at the slower growing *minus* ends of microtubules proximal to a centrosome matrix will be assumed positive. As will be noted in Section 1.3, the assumption of a net negative charge on centrosomes is consistent with experiment. However, positive charge at the minus ends of microtubules will induce negative charge on an electrically neutral centrosome matrix area adjacent to the free minus ends of the microtubules comprising the astral, polar, and kinetochore microtubule bundles, obviating the need to assume negative charge on centrosome matrices for some cell types. Calculations will be carried out in Chapters 3 and 4 for electrostatic force generation between positively charged microtubule minus ends and negatively charged centrosome matrices.

Similarly, kinetochores may manifest positive charge at pole-facing surfaces. Evidence for this comes from the presence within kinetochores of highly basic molecules in the Dam1 complex. In particular, the isoelectric points of Dam1p, Duo1p, and Spc34p are 9.97, 10.76, and 8.6, respectively. Significantly, experiments have revealed that the microtubule binding module of the Dam1 complex involves these three molecules; acidic proteins Ask1p, Spc19p, and Dad2p fail to bind [Westermann *et al.*, 2005].

Kinetochore molecules self-assemble onto highly condensed, negatively charged DNA at centromeres [Alberts *et al.*, 1994b], indicating that kinetochores may exhibit positive charge. This is an example of an important aspect of electrostatic interactions within cells: namely their longer range compared to other intracellular molecular interactions and the resulting capacity of electrostatic force to organize molecules and structures within cells. As will be discussed shortly, quite apart from the ability of microtubules to extend electrostatic interactions over cellular distances, the range of electrostatic fields within the cytosol itself is longer than ordinary counterion screening considerations would dictate. This will be seen throughout the present work to have important implications for a number of mitotic events.

Analogous to the situation for induced negative charge on a centrosome matrix, it may not be necessary to assume net positive charge on kinetochores since the negatively charged free plus ends of kinetochore microtubule bundles will induce positive charge on kinetochores. This possibility is considered in Chapter 4. The calculations in Chapters 3 and 4 demonstrate that the magnitude of the motive force for poleward motion of chromosomes is sufficient given either permanent or induced charge on centrosome matrices and kinetochores.

1.3   Spindle assembly and dynamics

It is reasonable to expect that the electric dipole nature of tubulin subunits greatly assists in their self-assembly into the microtubules of the asters and spindle. Thus we may envision that electrostatic fields organize and align the electric dipole dimer subunits, thereby facilitating their assembly into the microtubules that form the asters and mitotic spindle [Gagliardi,

2002b]. This self-assembly would be aided by significantly reduced counterion screening due to layered water adhering to the net charge of the dipolar subunits. Such water layering to charged proteins has long been theorized [Jordan-Lloyd and Shore, 1938; Pauling, 1945] and has been confirmed by experiment [Toney *et al.*, 1994]. Additionally, as will be described in Chapter 3, layered water between sufficiently close charged proteins has a dielectric constant that is considerably reduced from the *bulk* value far from charged surfaces, further increasing the tendency for an electrostatic assist to aster and spindle self-assembly. The question of what is meant by "sufficiently close" charged protein surfaces as well as the reduction in the dielectric constant between such surfaces will be addressed in Chapter 3.

The combination of these two effects (or conditions) – water layering and reduced dielectric constant – can significantly influence cellular electrostatics in a number of important ways related to cell division. It will be convenient in this book to characterize gaps between charged surfaces within cells that allow these two effects to significantly enhance electrostatic interactions as *critical separations* or *critical distances* (see Chapter 3). Thus these conditions would be expected to significantly increase the efficiency of microtubule self-assembly in asters and spindles by (1) allowing electrostatic interactions over greater distances than Debye screening dictates, and (2) increasing the strength of these interactions by an order of magnitude due to a corresponding order of magnitude reduction in the cytosolic dielectric constant (discussed in Chapters 2 and 3) between charged protein surfaces separated by critical distances or less.

As we will see in subsequent chapters, these two effects for charged surfaces at close range could also have important consequences regarding force generation for chromosome motions and other mitotic events.

The aster's pincushion-like appearance is consistent with electrostatics since electric dipole subunits will align radially outward

about a central charge with the geometry of the resulting config-
uration resembling the electric field of a point charge. From this
it seems reasonable to assume that the pericentriolar material,
the *centrosome matrix* within which the microtubule dimer dipolar subunits assemble in many cell types to form an aster [Joshi
et al.,1992], carries a net charge. This agrees with observations
that the microtubules appear to start in the centrosome matrix
[Wolfe, 1993a]. One may assume that the sign of this charge
is negative [Gagliardi, 2002a; Gagliardi, 2002b]. This assumption is consistent with experiments [Heald *et al.*, 1996] revealing
that mitotic spindles can assemble around DNA-coated beads
incubated in *Xenopus* egg extracts. The phosphate groups of
the DNA will manifest a net negative charge at the pH of this
experimental system. Recently, centrosomes have been shown
to have a net negative charge by direct measurement [Hormeño
et al., 2009].

As mentioned above, measurements have shown that the pH_i of
many cells rises to a maximum at the onset of mitosis and subsequently decreases throughout cell division. This could account
for the efficient self-assembly of the spindle during prophase,
when microtubule polymerization and microtubule organizing
center nucleation is favored because of the greater expression of
negative charge on centrosome matrices and tubulin dimers due
to the higher pH_i at this time. As indicated above, one possible
aspect of this would be the greater length of negatively charged
C-termini extending outward from microtubules contributing to
an increased probability for microtubule growth due to an increased probability for electrostatic interactions between tubulin
dimers in the higher pH_i during prophase.

An electrostatic component to the biochemistry of the microtubules in assembling asters is consistent with experimental observations of pH effects on microtubule assembly [Schatten *et
al.*, 1985], as well as the sensitivity of microtubule stability to
calcium ion concentrations [Weisenberg, 1972; Borisy and Olmsted, 1972]. Thus it would seem reasonable to assume that, over

distances consistent with the modified counterion screening discussed above, the electrostatic nature of tubulin dimers would allow tubulin dimer microtubule subunits (1) to be attracted to and align around charge distributions within cells – in particular, as mentioned above, around centrosomes – and (2) to align end to end and laterally, facilitating the formation of asters and mitotic spindles.

The motive force for the migration of asters and assembling spindles during prophase can also be addressed in terms of nanoscale electrostatics. As a consequence of the negative charge on the free plus ends of microtubules at the periphery of the forming asters and half-spindles, they would be repelled electrostatically from each other and drift apart. Specifically, as microtubule assembly proceeds, a subset of the negatively charged microtubule free ends at the periphery of one of the growing asters/forming half-spindles would mutually repel a subset of the negatively charged free ends at the periphery of the other, causing them to drift apart as assembly of their microtubules continues [Gagliardi, 2002b].

As discussed above, because of significantly reduced screening and the low dielectric constant of layered water adhering to the charged free ends of tubulin dimers, the necessary attraction and alignment of tubulin during spindle self-assembly would be enhanced by the considerably increased range and strength of the electrostatic attraction between oppositely charged regions of the tubulin dimers. Similarly, the mutually repulsive electrostatic force between a subset of like-charged plus ends of interacting microtubules from opposite half-spindles in the growing mitotic spindle would be expected to be significantly increased in magnitude and range. Thus mutual electrostatic repulsion of the negatively charged microtubule free plus ends distal to centrosomes in assembling asters/half-spindles could provide the driving force for their poleward migration in the forming spindle [Gagliardi, 2002b]. A subset of interacting microtubules in a small portion of a forming spindle is depicted in Figure 1.3.

It is important to note that interacting microtubules can result from either growing or shrinking microtubules but polymerization probabilities will dominate during prophase.

Interacting Microtubules

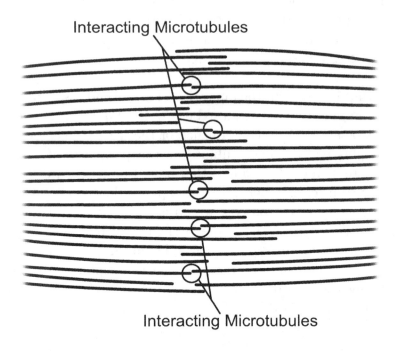

Interacting Microtubules

Fig. 1.3. A subset of interacting microtubules in a small portion of a forming mitotic spindle. The free plus ends of interacting microtubules within a few nanometers are mutually repelling. Protofilament curling of disassembling microtubules is not shown on this scale.

This process would continue until growing half-spindles drift as far apart as possible, establishing the cell poles.

1.4 Summary

In light of the large body of experimental information regarding mitosis, the complexity and lack of unity of models for the various events and motions gives, at least to this observer, reason to believe that approaching mitosis primarily within the molecular biology paradigm is flawed. This book offers a different approach based on nanoscale electrostatic interactions.

It seems clear that cellular electrostatics involves more than the traditional thinking regarding counterion screening of electric fields and the resulting unimportance within cells of the second most powerful force in nature. The reality may be that the evidence suggests otherwise, and that the resulting enhanced electrostatic interactions are more robust and act over greater distances than previously thought. One aspect of this is the ability of microtubules to extend the reach of electrostatic force over cellular distances; another lies in the reduced counterion scееening and dielectric constant of the cytosol between charged protein surfaces.

High pH_i during prophase favors spindle assembly. This includes greater electrostatic attractive forces between tubulin dimers as well as increased repulsive electrostatic interactions driving poleward movements of forming half-spindles.

Changes in microtubule dynamics are integral to changes in the motions of chromosomes during mitosis. These changes in microtubule dynamics can be attributed to an associated change in intracellular pH (pH_i) during mitosis. In particular, a decrease in pH_i during mitosis may act as a master clock controlling microtubule disassembly/assembly probability ratios by altering the electrostatic interactions of tubulin dimers. This, in turn, could determine the timing and dynamics of post-attachment mitotic chromosome motions. The possible electrostatic consequences of a subsequently decreasing pH_i on mitotic motions and events will be discussed in chapters to follow.

Chapter 2

Electrostatics in Nuclear Envelope Breakdown

2.1 Mechanical equilibrium of a nuclear envelope

The nuclear envelope is a double membrane composed of two lipid bilayers separated by a gap of about 40 nm, known as the perinuclear space. The perinuclear space surrounds the nucleus and is continuous with the lumen (open spaces) of the endoplasmic reticulum. The nuclear envelope has an electron-dense layer, the *nuclear lamina*, lying on the nucleoplasmic side of its inner membrane. The nuclear lamina consists of a 10 to 20 nm thick fibrous network of interconnected protein subunits called nuclear lamins. There appears to be some consensus that nuclear envelope breakdown is often preceded by phosphorylation of the nuclear lamins, resulting in nuclear lamina depolymerization and subsequent fragmentation of the nuclear envelope [Marshall and Wilson, 1997]. However, the nuclear envelope can remain intact even after nuclear lamina depolymerization occurs [Stick and Schwartz, 1983; Newport and Spann, 1987]. Membrane tears, induced by microtubules from the forming spindle, have also been associated with nuclear envelope breakdown [Beaudouin *et al.*, 2002; Salina *et al.*, 2002], but nuclear envelope breakdown still occurs in cells with severely compromised microtubule networks. In spite of the appearance of these and other recent models for nuclear envelope breakdown, the mechanisms controlling this process have remained somewhat elusive.

Previously, membranes of the nuclear envelope were thought to undergo a disassembly process to produce numerous vesicles (membrane-bound sacs) that become dispersed in the cytoplasm [Gerace and Blobel, 1980]. But observation of mitotic cells expressing fluorescent envelope proteins has revealed that these proteins move freely throughout a single, continuous endoplasmic reticulum membrane system, with no indication of vesicular intermediates [Ellenberg et al., 1997; Terasaki et al., 2001].

Mammalian cells are known to have bound negative electrical charge at their surfaces. This area surface charge density σ (C/m^2) is due largely to the ionized carboxyl groups of sialic acid residues [Seaman and Cook, 1965]. The net negativity is primarily a function of the density of these anionic groups fixed at the cell surface, but it also depends on the pH and ionic strength of the surrounding medium [Tamura et al., 1982]. Sialic acids have also been found on the nuclear envelope [Warren et al., 1975; Saito et al., 2002]. Other acidic groups may reside on the cell surface, and some cells have anionic groups associated with RNA at their peripheries [Weiss, 1969; Mayhew, 1966].

The hemispherical shell (Figure 2.1), representing half of the double membrane, is in equilibrium under the action of a uniform surface tension acting in the $-x$ direction from the rest of the membrane, and surface forces acting perpendicularly outward from the surface everywhere over the hemisphere. The surface forces arise from both the pressure difference across the membrane and a surface electrical force per unit area of the membrane (membrane electrostatic stress). These forces may be visualized if one imagines a charged balloon in equilibrium under (a) the surface tension forces due to the elastic deformation of the rubber, (b) the pressure difference Δp between the inside and outside of the balloon, and (c) an electrostatic stress due to the mutual repulsion of like charges fixed to the surface.

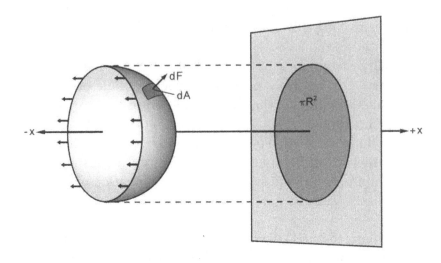

Fig. 2.1. Equilibrium of a nuclear envelope. The model spherical half nuclear envelope is in equilibrium under surface tension acting to the left and electrostatic stress plus pressure difference components acting to the right.

The use of a spherical shell is clearly an idealization. However, the analysis presented here is valid for a large class of ellipsoids of revolution, with the mathematics being far simpler for a spherical geometry.

The static mechanical equilibrium of the model half nuclear envelope of radius R and surface tension γ depicted in Figure 2.1 is considered in Appendix D where it is shown that

$$4\gamma/R = \Delta p + \sigma^2/2\varepsilon . \qquad (2.1)$$

The *bulk* phase value of ε for the cytosol (cytoplasmic fluid) of cells may be taken as approximately $71\,\varepsilon_0$, where ε_0 is the permittivity of free space, 8.85 pF/m. The room temperature bulk permittivity for water is $80\,\varepsilon_0$; the value $71\,\varepsilon_0$ incorporates corrections for the temperature and ionic depression of the dielectric constant of the cytosol [Hasted *et al.*, 1948]. It is well established in electrochemistry [Bockris and Reddy, 1977] that the permittivity of the first few water layers outside a charged surface is an order of magnitude smaller than that of the bulk phase. The Debye shielding of the electric field begins just beyond the water layers.

The effective permittivity of water as a function of distance from a charged surface has been determined by atomic force microscopy [Teschke et al., 2001] to increase monotonically from 4–$6\,\varepsilon_0$ at the surface to $78\,\varepsilon_0$ at a distance of 25 nm from the interface. This experiment was carried out with mica, which is known to have a surface charge density that varies from 1 to 50 mC/m^2, in the same range as biological surfaces [Pashley, 1981; Heinz and Hoh, 1999]. Thus, since the corrected bulk phase value of ε for the cytosol is less than $80\,\varepsilon_0$, the above experimental results indicate that a conservative value for ε in the cytosol just outside the charged nuclear envelope can be taken as $10\,\varepsilon_0$.

Studies have shown that electrostatic stress does manifest itself in membrane equilibrium. After a series of experiments on cell deformability, Weiss [Weiss, 1968] concluded that terminal sialic acids contributed to the mechanical properties of the cell or plasma membrane through electrostatic repulsion between their own and other anionic groups. The cells in this study became more deformable after incubation with neuraminidase, which removes the negatively charged sialic acid residues from the cell surface. In animal cells, the terminal sialic acids are attached to membrane proteins that are firmly anchored in the lipid bilayer. It would therefore seem reasonable to assume that repulsive interactions between these groups are primarily responsible for the observed decrease in the deformability of cells with more surface negative charge.

As mentioned above, the net negativity of bound charge is a function of the charge density σ of the bound anionic groups, as well as the pH and ionic strength of the surrounding medium

[Seaman and Cook, 1965; Tamura *et al.*, 1982]. If Δp were larger than $\sigma^2/2\,\varepsilon$, removal of a significant portion of surface charge by incubation with neuraminidase would not have been consistent with the observed increase in the deformability of these cells [Weiss, 1968]. We may therefore assume that the Δp term is at most comparable in magnitude to the $\sigma^2/2\,\varepsilon$ contribution to membrane stress. Accordingly, it should be possible to obtain a good approximation for the equilibrium of a half nuclear envelope from (2.1) written as follows:

$$\sigma^2/2\,\varepsilon = 4\,\gamma/R. \tag{2.2}$$

The range of values of γ may be taken as 0.1 to 1.0 dyne/cm (10^{-4} to 10^{-3} N/m) [Giese, 1968]. These values, when substituted into (2.2), using $R = 5$ μm and $\varepsilon = 10\,\varepsilon_0 = 88$ pF/m, give a range of values for σ from 0.12 to 0.38 mC/m². Values of σ in this range will be sufficient to exert an electrostatic stress comparable to nuclear envelope surface tension. Experimental values for the surface charge density of plasma and artificial lipid membranes range from 0.4 to 160 mC/m² [Segal, 1968; Fettiplace *et al.*, 1971].

2.2 Equilibrium of membrane charge clusters

The results of this section will be applicable to the question of the local mechanical stability of a membrane charge cluster against the tendency to separate from other charge clusters under an increased electrostatic stress. These charge clusters may be envisioned as randomly distributed islands of charge separated by areas of uncharged membrane. This picture is consistent with a *raft* structure with anchored proteins [Anderson and Jacobson, 2002]. The discovery of membrane rafts has recently shed more light on heterogeneity in biomembrane composition. It has been known for some time that the nuclear envelope and endoplasmic reticulum do not disintegrate down to the molecular level; rather, they disassemble into membrane fragments of many molecules each, allowing for a rapid and efficient reassembly in the nuclei and endoplasmic reticulum of the daughter cells [Cleveland, 1953].

A local charge cluster with area charge density σ_c on a membrane fragment of the nuclear envelope, depicted as a spherical cap of area ΔA in Figure 2.2, will be in equilibrium if the $-x$ components of the surface tension forces summed around the periphery of the cluster, $2\pi y\,\gamma\cos\alpha$, equals the total force in the $+x$ direction, $(\sigma_c^2/2\,\varepsilon)\Delta A + (p_1 - p_2)\,\Delta A$. Neglecting the Δp term, as before, the equilibrium relation for a charge cluster of the nuclear envelope is

$$(\sigma_c^2/2\,\varepsilon)\Delta A = 2\pi y\,\gamma\cos\alpha\,, \tag{2.3}$$

where γ and $\cos\alpha$ are the values at the membrane fragment and σ_c is the charge density on the charge cluster.

Since the radius R of the spherical cap representing the charge cluster in a typical cell is proportionately much larger than that shown (Figure 2.2), we have $y \ll R$, $\Delta A = \pi y^2$, and

$$(\sigma_c^2/2\,\varepsilon)\pi y^2 = 2\,\pi\gamma\,y\cos\alpha. \tag{2.4}$$

Thus, $\sigma_c^2/2\,\varepsilon = 2\gamma\cos\alpha/y$, and since $y = R\cos\alpha$ for $y \ll R$, we obtain

$$\sigma_c^2/2\,\varepsilon = 2\gamma/R. \tag{2.5}$$

The equilibrium of a membrane charge cluster is therefore governed (except for a factor of 2) by the same equation that was derived earlier for the equilibrium of the double membrane half nuclear envelope; however, this development highlights the strong tendency for membrane fragmentation due to the geometry effect. The easiest way to appreciate the effect of geometry is to consider the analogous situation of a circus tightrope walker. In that case, the tension in the cable is much greater than the weight of the performer because the force causing the tension (the person's weight) is nearly perpendicular to the cable.

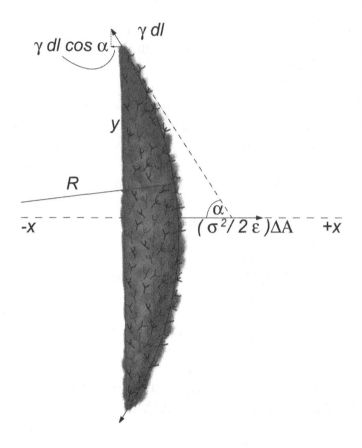

Fig. 2.2. Equilibrium of a membrane charge cluster.
The charge cluster is depicted in equilibrium under surface tension components acting to the left and pressure difference plus electrostatic stress components directed to the right. The radius R of the cluster is shown much smaller than actual for clarity.

Since electrostatic stress acts perpendicularly to a charge cluster, and the geometry is otherwise similar, any increase in electrostatic stress would require a much greater increase in membrane surface tension in order to maintain equilibrium. As will be discussed later, given the heterogeneity of biomembranes, increased electrostatic stress could be significant for the breakup of nuclear envelope charge clusters into membrane fragments of many molecules, as observed experimentally.

The range of values of γ may be taken as 0.1 to 1.0 dyne/cm $(10^{-4}$ to 10^{-3} N/m) [Giese, 1968]. These values, when substituted into (2.5), using $R = 5 \ \mu$m and $\varepsilon = 10 \ \varepsilon_0 = 88$ pF/m, give a range of values for σ_c from 0.085 to 0.27 mC/m^2. Values of σ_c in this range will be sufficient to exert electrostatic stress comparable to surface tension on charge clusters at the nuclear envelope. As mentioned above, experimental values for the surface charge density of plasma and artificial lipid membranes range from 0.4 to 160 mC/m^2.

2.3 Electrostatics in nuclear envelope disassembly

Cell electrophoresis studies have shown that there is a 12-18 % increase in plasma membrane negativity just prior to cell division [Mayhew, 1966; Mayhew and O'Grady, 1965]. This finding is consistent with an observed increase of 50 % in the whole cell content of sialic acid during mitosis [Glick et al., 1971]. We have

$$\frac{\Delta(\sigma^2)}{\sigma^2} \approx 2\Delta\sigma/\sigma, \qquad (2.6)$$

and an increase of 15 % in cell membrane negativity corresponds to an approximate 30 % increase in cell surface electrostatic stress. Similar increases may be expected for the membranes of the nuclear envelope, endoplasmic reticulum, and other endomembranes.

Experimental studies have shown that intracellular pH (pH$_i$) is highest during prophase in a number of cell types [Steinhardt and Morisawa, 1982]. In addition, as mentioned above, the whole cell content of membrane bound sialic acid is significantly increased at this time. These effects combine to cause a greater expression of net negative charge on the nuclear envelope and other membranes, including the plasma membrane. With respect to the cell surface, an increased electrostatic stress on the plasma membrane could explain the common observation that cells typically assume a more spherical shape ("round out") during mitosis.

The most immediate contribution to increased nuclear envelope net negative charge, with a resulting further increase in nuclear envelope electrostatic stress, may result from the timely release of Ca^{2+} from nuclear envelope and endoplasmic reticulum stores just prior to nuclear envelope breakdown [Silver et al., 1998; Silver et al., 1996; Silver, 1989; Stricker, 1995; Poenie et al., 1985].

This transient elevation in Ca^{2+} concentration can appear minutes before nuclear envelope breakdown, be highly localized in the perinuclear region, and last for short periods of time (100 ms or less) [Silver et al., 1998; Silver, 1994].

The calculations in the previous section indicate that, even without these Ca^{2+} releases, membrane charge density is large enough to be significant in the equilibrium of cellular membranes. As also discussed above, an increase in net charge – with a resulting elevated electrostatic stress – requires a considerably larger increase in membrane surface tension in order to maintain equilibrium of the charge clusters.

It would therefore seem that increased electrostatic stress, culminating in the observed release of Ca^{2+} just prior to nuclear envelope breakdown, could account for the mechanism as well as the timing of nuclear envelope fragmentation.

Although not discussed here, the same mechanism may also apply to endoplasmic reticulum and other endomembranes. Models for nuclear envelope breakdown in the current literature do not explicitly address how the membranes of the nuclear envelope disassemble into membrane fragments of many molecules each, nor do they appear to have the possibility of accounting for the breakdown of other endomembranes during mitosis within a unified model.

2.4 Energy considerations in nuclear envelope disassembly

A necessary condition for the spontaneous transition from the parent nucleus to two daughter nuclei, assuming no additional energy input (i.e., a purely electrostatic mechanism), is that the total membrane surface energy (electrostatic plus surface tension) of the system comprised of the two daughter nuclei be less than the energy of the parent nucleus. The electrostatic self-energy U_e of a sphere of radius R and surface charge Q is given by $Q^2/8\pi\varepsilon R$ [Griffiths, 1999a; Appendix A].

The surface energy due to membrane surface tension is obtained from the relation $U_s = 8\pi\gamma R^2$. It is convenient to motivate this equation by a dimensional argument. As we have seen, the dimensions of γ are force (Newtons) per meter (N/m) in SI units. Dimensionally, N/m is equivalent to energy (in Joules) per square meter (J/m^2). Thus the total surface area for the double membrane of the nuclear envelope $8\pi R^2$, multiplied by the energy per unit area γ gives the surface energy from surface tension, $8\pi\gamma R^2$.

The total membrane-associated energy of the parent nucleus of radius R is the sum of three contributions,

$$U_0 = \frac{Q^2}{8\pi\varepsilon R} + 8\pi\gamma\, R^2 + \frac{1}{2} \int_R^\infty \rho\phi\, d\tau, \qquad (2.7)$$

where ρ and ϕ are the ionic charge density and electric potential in the cytosol surrounding the nucleus. The first term expresses the electrostatic self-energy of a charged spherical membrane, and the last term represents the energy stored in the positively charged Debye "ionic atmosphere" U_{0a} (see Appendices A and E) surrounding the negatively charged parent nucleus.

Similarly, the total energy of the two nuclei in the nascent daughter cells (the daughter nuclei), each of radius R_1 and charge Q_D, is

$$U_1 = 2 \left[\frac{Q_D^2}{8\pi\varepsilon R_1} + 8\pi\gamma R_1^2 + \frac{1}{2} \int\limits_{R_1}^{\infty} \rho\,\phi\,d\tau \right] + \frac{Q_D^2 e^{-(d-R_1)/D}}{4\pi\varepsilon d(1 + R_1/D)} , \qquad (2.8)$$

where $d \geq 2R_1$ is the distance between the centers of the daughter nuclei, and the last term in this equation is the interaction energy of the two daughter nuclei. Due to Debye shielding over a distance such as that between the daughter nuclei, the interaction term will be extremely small, and we will neglect it in the calculation to follow. The magnitude of this interaction term is of the same order as that which we will calculate for the energies of the Debye ionic atmospheres (often called "Debye clouds"). The Debye cloud energies will be shown to be several orders of magnitude smaller than the other terms. As is the case with the interaction energy term, the exponential decay of electrostatic fields in an ionic medium is responsible for the small values of these energies.

A necessary condition for the spontaneous splitting of a parent nucleus of radius R and charge Q into two daughter nuclei of radius R_1 and charge Q_D is $U_0 > U_1$. Therefore, from (2.7) and (2.8) without the interaction term, we have

$$\frac{Q^2}{16\pi\varepsilon R} + 4\pi\gamma R^2 + \frac{1}{4} \int\limits_{R}^{\infty} \rho\,\phi\,d\tau > \frac{Q_D^2}{8\pi\varepsilon R_1} +$$

$$8\pi\gamma R_1^2 + \frac{1}{2} \int\limits_{R_1}^{\infty} \rho\,\phi\,d\tau \qquad (2.9)$$

The calculation in Appendix E shows that the energies of the ionic atmosphere energy terms are four orders of magnitude smaller than the other terms in the above equation. From (E.10) in Appendix E we also have that

31

$$\frac{\sigma^2}{2\varepsilon}\left(1 - \frac{2}{\beta^3}\right) > \frac{2\gamma}{R}\left(\frac{2}{\beta^2} - 1\right), \qquad (2.10)$$

where $\beta = R/R_1$, $Q = 4\pi\sigma R^2$, $R_1 = R/\beta$, and $Q_D = Q/\beta^2$. Previously, we had obtained the equilibrium condition (2.2):

$$\sigma^2/2\varepsilon = 4\gamma/R.$$

Thus, from (2.2) and (2.10), two necessary conditions regarding the energetics of nuclear envelope disassembly and subsequent reassembly based on electrostatic stress and surface tension are

$$\sigma^2/2\varepsilon > 4\gamma/R, \qquad (2.11)$$

and

$$\left(\frac{2}{\beta^2} - 1\right) < 2\left(1 - \frac{2}{\beta^3}\right). \qquad (2.12)$$

Assuming that membrane fragments from the parent double membrane nucleus will furnish the bulk of the fragments that are reassembled into the two double membrane daughter nuclei, we have

$$2(4\pi R^2) \leq 2[2(4\pi R_1^2)]. \qquad (2.13)$$

Simplifying,

$$R \leq \sqrt{2}R_1. \qquad (2.14)$$

In Section 2.1, it was argued that it is not difficult to satisfy (2.11) for known experimental values of membrane charge density. Since β is a positive number, (2.11), (2.12), and (2.14) require that the ratio of parent to daughter nuclear diameters β lie between 1.3 and 1.4.

2.5 Electrostatics in nuclear envelope reassembly

Experimental studies of a number of cell types show that pH_i has decreased considerably from its prophase peak by the onset of metaphase, and continues to decrease through the end of mitosis [Steinhardt and Morisawa, 1982]. After peaking earlier in mitosis, the whole cell sialic acid content also decreases through mitosis [Glick *et al.*, 1971]. These changes would cause the net negative charge on membrane fragments to decrease, allowing the fragments to approach more closely because of their reduced mutual electrostatic repulsion. This close approach would be essential for their subsequent self-assembly into the nuclear envelopes of the daughter cells.

The tendency for membrane fragments to approach close enough for membrane reassembly biochemistry to occur can be determined from a comparison of the average thermal energy of a membrane fragment with the repulsive electrostatic interaction energy of two membrane fragments. From the equipartition theorem of classical statistical mechanics, the average (rotational and translational kinetic) energy $< E >$ of an entity in thermal equilibrium with a system at absolute temperature T is given by the expression $< E >= 3\,kT$, where k is Boltzmann's constant. For the electrostatic interaction energy (see Appendix A) between two like charged spherical membrane fragments of radius a separated by a very thin cytosol layer of permittivity ε, we have

$$\frac{q^2}{4\,\pi\,\varepsilon\,(2a)}\,, \qquad (2.15)$$

where the charge q on the two membrane fragments is assumed equal, and $\varepsilon = 10\,\varepsilon_0$, as discussed in section 2.1.

For membrane reassembly to occur, the membrane fragments must approach each other's "surfaces." In order for one spherical fragment of radius a to reach the surface of another, it will need an amount of thermal energy given by

$$3kT = \left(\frac{1}{N}\right)\frac{q^2}{8\pi\varepsilon\,a}\,, \qquad (2.16)$$

where N is a small integer (between 1 and 10) specifying the fractional value of the electrostatic repulsive energy assigned to thermal energy, $3kT$. By definition, $q = 4\pi\, a^2 \sigma_c$, and we may write this as

$$kT = \frac{2\pi \sigma_c^2\, a^3}{3\varepsilon\, N}.$$

(2.17)

Solving this for $\sigma_c \sqrt{a^3/N}$, we have

$$\sigma_c \sqrt{a^3/N} = \sqrt{\frac{3\,\varepsilon\, kT}{2\pi}}.$$

(2.18)

Since the numerical value of the right-hand side of this equation is 4.7×10^{-16}, we may examine various possible combinations of the quantities on the left-hand side. One set of values consistent with this might be $N=4$, $a = 40$ nm, and $\sigma_c = 0.12$ mC/m^2. The values in this solution set and a number of other possible neighboring sets satisfying the above relationship are quite reasonable.

2.6   Summary

Experimentally observed increases in whole cell sialic acid content and intracellular pH during prophase, followed by an observed release of free calcium from nuclear envelope and endoplasmic reticulum stores, will significantly enhance the expression of negative charge on sialic acid residues of the nuclear envelope, providing sufficient electrostatic energy for nuclear envelope breakdown. Since terminal sialic acids are attached to membrane proteins that are firmly anchored in the lipid bilayer, the observed disassembly of the nuclear envelope is consistent with electrostatic repulsion between membrane charge clusters, which may tear apart under the influence of increased electrostatic charge.

Experimental observations regarding the mechanical properties of the plasma membrane show that electrostatic stress manifests itself in ways consistent with this scenario.

It is difficult to envision a purely biochemical process that would result in the nuclear envelope breaking into fragments of many molecules each. Models for nuclear envelope breakdown in the current literature do not address this problem, nor do they attempt to link nuclear envelope breakdown to the related problems of the breakdown of the endoplasmic reticulum and other endomembranes during cell division.

As is the case with the mechanical equilibrium approach, energy considerations indicate that experimentally measured values of membrane charge are sufficient to effect nuclear envelope fragmentation.

The observed lowering of both intracellular pH and whole cell sialic acid content during late anaphase and telophase is consistent with a decreased manifestation of net negative charge on internal membrane fragments at that time. These decreases could shift the balance of thermal energy versus electrostatic repulsive energy in favor of thermal energy, allowing the closer approach of membrane fragments necessary for reassembly to occur in nascent daughter cells.

Chapter 3

Electrostatic Force in Poleward Chromosome Motions

3.1   Introduction

A number of experiments have revealed that poleward motion of chromosomes proceeds by kinetochore microtubule disassembly in the vicinity of kinetochores [Gorbsky *et al.*, 1987; Mitchison *et al.*, 1986] or poles [Mitchison and Salmon, 1992; Maddox *et al.*, 2002; Zhang and Chen, 2003]. In the present work, poleward motion of chromosomes is addressed in terms of an electrostatic microtubule disassembly force. The factors: (1) an intracellular pH decrease from a peak at prophase through cell division, (2) an electrostatic component to microtubule disassembly/assembly probability ratios, (3) a nanoscale electrostatic microtubule disassembly force acting at kinetochores and centrosome matrices, and (4) nanoscale electrostatic microtubule assembly forces acting between astral microtubule plus ends and like charged chromosome arms, all working in conjunction with microtubule dynamics, make it possible to address the motive force – and timing – for chromosome motions throughout post-attachment prometaphase, metaphase, and anaphase in terms of electrostatics.

Electrostatic disassembly as well as assembly forces operate during post-attachment prometaphase, metaphase, and anaphase chromosome motions. The relative dominance of each is what changes. It is proposed in this work that pH_i decreases during prometaphase and metaphase may control the microtubule disassembly/assembly probability ratio and the resulting chromosome motions during these stages of mitosis. Specifically, chromosome post-attachment motions are viewed as determined primarily by changes in microtubule dynamics caused by a steadily decreasing pH_i. It is further proposed that the motive force for these movements is due to poleward electrostatic microtubule disassembly forces at kinetochores and poles acting in conjunction with antipoleward electrostatic microtubule assembly forces at chromosome arms. We now discuss the nature of an electrostatic poleward-directed microtubule disassembly force. Antipoleward electrostatic microtubule assembly force will be addressed in Chapter 5.

3.2 Electrostatic microtubule disassembly force at cell poles

The decrease in pH_i from a peak at prophase through mitosis that is observed for a number of cell types may be central to understanding the events of mitosis. A decrease in the microtubule assembly to disassembly probability ratio through metaphase is consistent with prophase microtubule net assembly as well as prometaphase and metaphase chromosome dynamics, which will be discussed in Chapter 5. The possible effect of a further decrease in pH_i during anaphase may be masked by the free calcium concentration ($[Ca^{2+}]$) increase that occurs in many cell types at the onset of anaphase-A. This situation will also be addressed in Chapter 5.

It is important to note that pH in the vicinity of the negatively charged exposed plus ends of microtubules will be even lower than the bulk pH_i because of the effect of negative charge at the free plus ends of the microtubules. This lowering of pH in the vicinity of negative charge distributions is a general result. Intracellular pH in such limited volumes is often referred to as *local* pH. As one might expect from classical Boltzmann statistical mechanics, the hydrogen ion concentration at a negatively charged surface can be shown to be the product of the bulk phase concentration and the Boltzmann factor $e^{-e\zeta/kT}$, where e is the electronic charge, ζ is the (negative) electric potential at the surface, and k is Boltzmann's constant [Hartley and Roe, 1940]. For example, for typical mammalian cell membrane negative charge densities, and therefore typical negative cell membrane potentials, the local pH can be reduced 0.5 to 1.0 pH unit just outside the cell membrane. Because of the negative charge at the plus ends of microtubules, a reduction of pH would be expected in the immediate vicinity of these free ends making the local pH influencing microtubule dynamics considerably lower, and a lower bulk pH_i would be accompanied by an even lower local pH.

The above discussion of lower local pH in the vicinity of negative charge distributions also applies to the cytosol region nearest a centrosome matrix (the *vicinal* cytosol), increasing the instability of microtubule minus ends just outside the centrosome boundary.

Experimental observations on poleward force generation at cell poles for post-attachment chromosome movements can be addressed in terms of electrostatics as follows. Microtubules invariably assemble or disassemble at their ends; that is, at some discontinuity in their structure. Furthermore, they are known to be in a constant condition of dynamic instability near the balanced state [Wolfe, 1993b]. From the discussion in Chapter 1, the net charge on the free ends of microtubules at a centrosome matrix is assumed to be positive. A γ-tubulin molecule, embedded in the fibrous matrix, takes the form of a ring from which a microtubule appears to emerge [Alberts, 1994c]. This would allow the electric field of the negatively charged γ-tubulin rings to draw the positively charged ends of microtubules into the centrosome matrix, with the resulting rapid change of the electric field just outside and across the outer boundary of the centrosome matrix destabilizing microtubules as they pass into the charge distribution.

Thus γ-tubulin rings may be regarded as forming a firmly anchored negative charge distribution near the surface of a centrosome matrix through which microtubules pass, disassembling in the passage, as depicted in Figure 3.1. As in the case for a kinetochore (discussed later in this chapter), the microtubules do not necessarily need to pass through the rings; rather, the rings provide a stable negatively charged volume distribution attracting non-penetrating microtubules to and into the centrosome matrix while exerting force on the microtubules.

As noted above, observations on a number of cell types have shown that disassembly of microtubules at spindle poles accompanies chromosome poleward movement. Accordingly, within the context of the present work, force generation at spindle poles for prometaphase post-attachment, metaphase, and anaphase-A poleward chromosome motions can be attributed to an electrostatic attraction between the positively charged free ends of disassembling kinetochore microtubules and a negatively charged centrosome matrix.

We now calculate the magnitude of the force produced in this manner by a non-penetrating microtubule at a centrosome matrix. Since the outer diameter of a centrosome matrix is considerably larger than the diameter of a microtubule, we may model it as a large, approximately planar slab with negative surface charge density of magnitude σ as depicted in Figure 3.1. From the well-known Debye-Hückel result for a planar charged surface with area charge density σ immersed in an electrolyte [Benedek and Villars, 2000c], we have for the electrostatic potential

$$\phi(x) = \frac{D\sigma}{\varepsilon}e^{-x/D} \, , \tag{3.1}$$

39

where D is the *Debye length* and x is the distance from the surface.

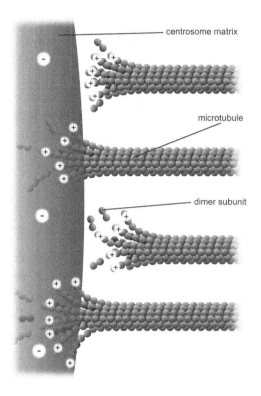

Fig. 3.1. Nanoscale electrostatic disassembly force at a centrosome. A poleward force results from an electrostatic attraction between positively charged microtubule free ends and an oppositely charged centrosome matrix. Only disassembling microtubules are depicted; assembling microtubules could also be momentarily attracted to a centrosome.

Appendix B contains an introduction to the Debye theory of counterion screening with a derivation of the electrostatic potential for the spherically symmetric case that is needed in Chapter

2. An entirely similar derivation for a planar charged surface results in (3.1). For a bulk dielectric constant of 71, the cytosol permittivity ε is $71\,\varepsilon_0$, where ε_0 is the permittivity of free space. The room temperature permittivity of water is $80\,\varepsilon_0$; the value of $71\,\varepsilon_0$ incorporates corrections for the temperature and ionic depression of the dielectric constant [Hasted et al., 1948] appropriate to the cytosol of mammalian cells.

The electric field $E(x)$, obtained from the negative gradient of the electrostatic potential, multiplied by the charge q gives the magnitude of the attractive force $F(x)$ between the charge q on a dimer subunit at the end of a protofilament and the centrosome. This results in

$$F(x) = q\,E(x) = -q\frac{\partial \phi(x)}{\partial x} = \frac{\sigma q}{\varepsilon(x)}e^{-x/D} \ . \qquad (3.2)$$

It is well established in electrochemistry [Bockris and Reddy, 1977] that the permittivity of the first few water layers outside a charged surface is an order of magnitude smaller than that of the bulk phase. As mentioned above, the effective permittivity of water as a function of distance from a charged surface has been determined by atomic force microscopy [Teschke et al., 2001] to increase monotonically from 4–$6\,\varepsilon_0$ at the interface to $78\,\varepsilon_0$ at a distance of 25 nm from the interface. The values of the dielectric constants $k_1(x)$ at distances of 1, 2, 3, and 4 nm from a charged surface were measured to be 9, 21, 40, and 60, respectively. As discussed in Chapter 1, layered water adhering to the net charge of proteins will significantly reduce counterion screening for small distances from the surface.

The interpolated values of $k_1(x)$ for separations between charged surfaces of up to 3 nm are are 5, 9, 9, and 5 for $x = 0, 1, 2$, and 3 respectively, where the charged surfaces are at $x = 0$ and $x = 3$ nm. The distance range 1 to 3 nm between charged surfaces is significant for the present calculation because 1 nm may be taken as the thickness of layered water adsorbed to each charged surface [Pauling, 1945; Pollack, 2001], and for charged surface separations up to 3 nm, counterion screening would be virtually eliminated. Thus for charged surfaces at separations of 3 nm or less, the exponential decay of the electric field from each surface reduces the electric field to a minimum of approximately 60 % of the maximum value at each surface and, to a good approximation, Debye screening may be neglected. For brevity in subsequent discussions, separations of 0 to 3 nm between charged surfaces will be designated as *critical separations.* The electrostatic potential energy ϕq between charged surfaces separated by 5 nm is near the upper limit of the effective range within which thermal energy is less than electrostatic energy for a charge q of one electron charge.

For critical separations, the expression for the force between a charged centrosome matrix surface at $x = 0$ and a charge q on the free minus end of a protofilament at a distance x from the surface may therefore be written

$$F(x) = \frac{\sigma q}{\varepsilon(x)} \ , \qquad (3.3)$$

where $\varepsilon(x) = k(x)\,\varepsilon_0$ is obtained from the interpolated experimental results for $k(x)$ referred to above, $\varepsilon_0 = 8.85$ pF/m and q is the charge on the protofilament free end. This equation may be obtained from (3.2) in the limit as $D \to \infty$, a condition that effectively eliminates counterion screening.

There are 13 protofilaments arranged circularly in a microtubule, with an axial shift of 0.92 nm for each protofilament as one moves around the circumference of a B lattice microtubule [Brown and Tuszyński, 1997]. For comparison with experimental values, a calculation of the maximum disassembly force per microtubule will be carried out for 3 protofilaments with their free ends at distances of 1, 2, and 3 nm from the centrosome matrix surface. The actual distribution for the distances of the free ends of 13 – disassembling (curling), and temporarily assembling (straight) – protofilaments would be considerably complicated, and it is probable that more than 3 protofilaments will interact with a centrosome matrix within critical distances.

As mentioned above, experimental values of surface charge density σ for biological surfaces range from 1 to 50 mC/m^2. Thus, we may sum the forces on protofilament free ends at distances of 1, 2, and 3 nm from the centrosome matrix using the above interpolated values of $\varepsilon(x)$ and a conservative value for σ of 10 mC/m^2. Carrying out this calculation with (3.3), we find that the electrostatic force sums to 23 n pN/MT (picoNewtons per microtubule), where $q = n\,e$, with e equal to the charge on an electron and n the number of electron charges at a protofilament free end. Comparing this value with the experimental range of 1–74 pN/MT [Alexander and Rieder, 1991] for the maximum tension force per microtubule, we have that $n = 0.04 - 3.2$ electron charges. This range of values compares favorably to experiments [Brown and Tuszyński, 1997; Tuszyński et al., 1998; Stracke et al., 2002], and the agreement represents a successful *ab initio* theoretical derivation of the magnitude of this force; however, this calculation is primarily intended to demonstrate that nanoscale electrostatic interactions are able to produce a force per microtubule within the experimental range.

We now proceed to a calculation of the electrostatic force on a kinetochore due to penetrating microtubules. It will be convenient to temporarily postpone a calculation of the force on a microtubule penetrating a centrosome matrix.

3.3 Electrostatic microtubule disassembly force at kinetochores

Experimental observations on force generation at kinetochores may also be addressed using electrostatics. As discussed in Chapter 1, kinetochore pole-facing surfaces may manifest positive charge. Recent work has implicated the Dam1 complex within kinetochores in coupling kinetochores to the ends of shortening kinetochore microtubules during mitosis. [See, for example, Grishchuk *et al.*, 2008a]. This is consistent with an electrostatic microtubule disassembly force at kinetochores because the highly positive molecules Dam1p, Duo1p, and Spc34p of the Dam1 complex will be attracted within critical nanometer distances to the highly negative plus ends of shortening microtubules. As mentioned in Section 1.2, experiments reveal that the microtubule binding module of the Dam1 complex involves these proteins [Westermann *et al.*, 2005].

A distance interaction for non-penetrating microtubules is necessary for efficient tracking and recoupling of kinetochores to microtubules throughout the complex motions of mitosis. It is important to recognize this, and any model of mitotic chromosome motility must account for it.

As we now discuss, force generation will continue for microtubules that penetrate kinetochores. Assuming a volume positive charge at kinetochore pole-facing surfaces, we may envision a mechanism for electrostatic force generation at kinetochores by penetrating microtubules. It has been accepted for some time that electron microscope studies show kinetochore microtubules running uninterupted between poles and kinetochores, terminating in the outer poleward-facing *plate* of the kinetochores [Rieder, 1982]. It has also been assumed that this kinetochore-microtubule association is the locus of force generation. As a result, ultrastructural studies of kinetochore-microtubule associations have concentrated on the microtubules that are apparently penetrating the outer plate of kinetochores, and possibly being pulled into kinetochores by interactions between kinetochores and their microtubules to produce poleward force.

These interactions have been described in more specific terms in the current literature, and may be classified as being based on motor molecules [Civelekoglu-Scholey *et al.*, 2006], or binding relationships between protofilaments and kinetochore molecules [Grishchuk *et al.*, 2008b; McIntosh *et al.*, 2007; Guimareas *et al.*, 2008].

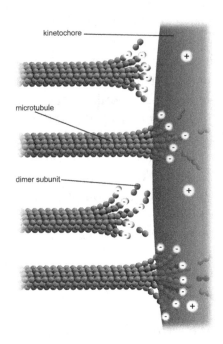

Fig. 3.2. Nanoscale electrostatic disassembly force at a charged kinetochore. A poleward force results from an electrostatic attraction between negatively charged microtubule free ends and an oppositely charged kinetochore.

Since kinetochore plate diameters are generally large compared to the diameters of microtubules, we may model the kinetochore microtubule interaction for penetrating microtubules by assuming an approximately planar slab of uniform positive charge density, with thickness a parallel to the x axis [Gagliardi, 2005], for the outer kinetochore plate interacting with negatively charged free ends of microtubule protofilaments, as depicted in Figure 3.2. The possibility of binding to molecules, or Dam1 rings around microtubules, is not assumed in favor of a more general approach wherein interacting kinetochore molecules comprise a firmly anchored positive charge distribution through which negative charge at or near the free plus ends of kinetochore microtubules are drawn.

Experiments have revealed that kinetochore-microtubule binding is dependent on electrostatic interactions mediated via positive charge in the disordered N-terminal 80 amino acid tail domain of Hec1 [Guimaraes *et al.*, 2008]. This result may be regarded as further support for an electrostatic interaction between known negative charge at and near the plus ends of microtubules and positive charge distributions at kinetochores as the "motor" for poleward force generation at kinetochores.

A standard result from an application of Gauss's law [Griffiths, 1999b] gives the following result for the electric field inside a large, uniformly charged slab of positive charge

$$E(x) = \rho x / \varepsilon_2 , \tag{3.4}$$

where ρ is the volume charge density, ε_2 is the dielectric permittivity of the slab, and $x = 0$ at the plane of symmetry in the center of the large rectangular slab. (Note that previously in (3.3), $x = 0$ at the right boundary of the centrosome matrix, as shown in Figure 3.1.) Making use of the uniform charge relation $\sigma = \rho a$, this result may be expressed in terms of the surface charge density σ as

$$E(x) = \sigma x / \varepsilon_2 a . \tag{3.5}$$

Electron microscope studies reveal that kinetochore microtubules penetrate only the outer (poleward-facing) plate on each kinetochore [Rieder, 1982]. The force on a protofilament of negative charge magnitude q at its free end a distance x from the outer plate's plane of symmetry is given by

$$F(x) = qE(x) = q\sigma x / \varepsilon_2 a . \tag{3.6}$$

At the left face of the slab (Figure 3.2), $x = -a/2$, and $E = -\sigma / 2\varepsilon_2$, and the force exerted in the negative x (poleward) direction on a kinetochore by a protofilament free end with negative charge of magnitude q at its free end located just inside the left face is $-\sigma q / 2 \varepsilon_2$.

The value of the dielectric constant k_2 for a kinetochore has not been established. Consistent with their open structures, a cytosol-saturated kinetochore or centrosome matrix would be expected to have a dielectric constant that is quite large, roughly midway between their *dry* values and that of cytoplasmic water [Schelkunoff, 1963]. As with a centrosome matrix, (1) the value for cytoplasmic water will dominate, and (2) the calculation is relatively insensitive to the precise dry value. From these considerations, k_2 can be taken as 30.

Using $k_2 = 30$ and the value $\sigma = 10$ mC/m^2 in carrying out a conservative calculation for a microtubule with 6 of the 13 protofilament ends – irrespective of protofilament curling – at an average distance $x = -a/4$ from the symmetry plane, $x = 0$ (where the force is 0), we find that the force on a penetrating microtubule sums to $90n$ pN/MT. Equating this result to the experimental range $1 - 74$ pN/MT, we find that $n = 0.01 - 0.82$ electron charges, again well within the experimental range.

An evaluation of the force on a kinetochore from a non penetrating microtubule within a critical distance of the kinetochore mirrors the previous calculation for a centrosome matrix. As mentioned above, it has generally been assumed that penetrating kinetochore microtubules are responsible for force generation. Consequently not much attention has been given to the possibility that kinetochore microtubules may be generating force in non-contact interactions such as those arising from electrostatics.

Force generation by nanoscale electrostatic non-contact interactions, primarily over critical separations, could cause other (previously force-generating) kinetochore microtubules in the bundle to penetrate the kinetochore, giving the illusion of contact force

47

generation by penetration. Importantly, as also noted above, forces acting at nanometer distances would seem to be essential for efficient microtubule reattachment and tracking to kineto-chores throughout mitosis, a feature that is not addressed by any of the current models for chromosome motility.

As in the case for poleward force generation at a centrosome, non-penetrating microtubules that disassemble in the region of high electric field gradient just outside the outer plate of a charged kinetochore also generate a poleward force, as depicted in Figure 3.2. Because of the similarity in geometry, a calcu-lation of the force per microtubule for non-penetrating micro-tubules at kinetochores will yield essentially the same result as the calculation at a spindle pole. Accordingly, a force calculation with (3.3) carried out with $\sigma = 10$ mC/m^2 also reproduces the experimental range 1–74 pN/MT as the nanoscale electrostatic microtubule disassembly force at a kinetochore for the experi-mentally verified charge range of 0.04 to 3.2 electron charges at a protofilament free end.

As mentioned earlier, given the electric dipole nature as well as the large overall net charge on C-termini of tubulin microtubule subunits, the electric field gradient over a critical separation distance within vicinal cytosol at a centrosome matrix or kine-tochore could act to destabilize the microtubules. In addition, the field gradient across either boundary may destabilize micro-tubules, further increasing the depolymerization probability for microtubules as force is generated, in agreement with observa-tion.

These gradients will now be quantified. To gain perspective (since electric fields and electric field gradients are less intu-itive) it will be convenient to calculate force gradients. For the electrostatic force gradient in the vicinal cytosol at a kineto-chore or centrosome matrix, we may consider a protofilament free end with one electron charge e starting at a distance of 1 nm where the dielectric constant is 9, then moving to the surface where the dielectric constant is 5 (see interpolated values of $k_1(x)$ given in Section 3.2). From $x = 1$ to 0, we have a force increase ΔF of $\sigma\,e/5\varepsilon_0 - \sigma\,e/9\varepsilon_0$. Evaluating the average force gradient, $\Delta F/\Delta x$ over the separation distance $\Delta x = 1$ nm, we have that $\Delta F/\Delta x = 166$ pN/nm. This is a rather large average force gradient. Similarly, the average gradient over 2 nm experienced by a protofilament free end with one electron charge penetrating a centrosome matrix or kinetochore boundary where the force changes from $\sigma\,e/5\varepsilon_0$ to $\sigma\,e/2\varepsilon_2$ is 160 pN/nm.

3.4 Penetrating microtubules at a centrosome

To complete the possibilities for poleward force production, we now consider the situation for microtubules penetrating a centrosome matrix. The value of the dielectric constant k_2 for a centrosome matrix has also not been established. As discussed above for a kinetochore, due to an open structure that allows cytoplasmic water intrusion, the large dielectric constant of water would strongly influence the overall dielectric constant of the centrosome matrix, leading to a value that is relatively insensitive to the dry value. As discussed previously, consistent with their open structures, a cytosol-saturated centrosome matrix or kinetochore would be expected to have a dielectric constant that is quite large, again roughly midway between their dry values and cytoplasmic water. Therefore, as with a kinetochore, (1) the value for cytoplasmic water will dominate, and (2) the calculation is relatively insensitive to the precise dry value. For simplicity, the conservative value $k_2 = 30$ will again be assumed.

In discussing the force calculation at a kinetochore, the possibility that molecules of the Dam1 complex can form rings around microtubules was not considered fundamental. Instead, a more general approach considered these molecules as contributing a structurally stable positive charge distribution that negatively charged microtubule plus ends can be attracted to and drawn into. Similarly, γ-tubulin rings may be viewed as a firmly anchored negative charge distribution through which the positively charged minus ends of kinetochore microtubules are drawn, generating poleward force associated with microtubule poleward flux.

As a result of these similarities, and the assumed approximate equality of k_2 for both a kinetochore and a centrosome matrix, a calculation of the poleward force per microtubule for penetrating microtubules at a centrosome matrix will yield a result identical to the above calculation at a kinetochore. As in all the calculations in this chapter, since the calculated range of n is well within the experimental range, it would seem reasonable to assume that moderate differences in k_2, the geometry, and other contributing factors would not be significant.

3.5 Summary

Given the known net negative charge at the plus ends of microtubules and the presence of highly basic molecules in kinetochores it is difficult to imagine there not being an attractive electrostatic poleward-directed force between the plus ends of kinetochore microtubules and kinetochores. Calculations of electrostatic force magnitudes for penetrating and non-penetrating microtubules within critical separations show that nanoscale electrostatic interactions are able to account for poleward-directed force production at both kinetochores and poles. The calculated maximum force per microtubule falls within the experimental range, and represents a successful *ab initio* derivation of the magnitude of this force. A simulation (see Section 4.4) supports this calculation.

In agreement with experiment [Nicklas, 1988], the calculations given in this chapter satisfy the requirement that the maximum tension force per microtubule be the same for all mitotic chromosome attachments. The approximate equality in the calculations for electrostatic force generation by non-penetrating or penetrating kinetochore microtubules at both poles and kinetochores, as well as parity between overall force generation at poles and kinetochores demonstrated here, is to be expected of any dynamical mechanism for poleward force generation. Models for chromosome motility must account for these parities.

Models for chromosome motility should also address the efficiency with which spindle microtubules maintain coupling to kinetochores and poles throughout the various movements during mitosis. Electrostatic force at nanometer distances between the free ends of a kinetochore microtubule and a kinetochore or centrosome matrix has this property.

Consistent with observation, spindle microtubules depolymerize while generating force at kinetochores and poles. Given the electrostatic nature of tubulin microtubule subunits, this can be understood in terms of the large electric field (and therefore force) gradients in critical distances within vicinal cytosol outside as well as across the boundaries of kinetochores and centrosome matrices.

Chapter 4

Induced Charge
in Poleward Motions

4.1   Introduction

Internal consistency within the present work requires that the charge on the plus free ends of microtubules proximal to kinetochores be negative. It is generally accepted that this is the case. Within the context of induced charge in the present chapter, either sign of the microtubule minus end net charge at the centrosome matrix is possible. However, for consistency with experiments indicating a net negative charge on centrosome matrices (see Section 1.3), and because of the observed electric dipole nature of tubulin dimer subunits comprising microtubules, it will be assumed that the net charge at the minus ends of microtubules proximal to a centrosome matrix is positive.

As noted previously, highly basic molecules in kinetochores indicate that they may manifest positive charge. However, as discussed above for induced negative charge on the centrosome matrix, it may not be necessary to assume a permanent positive charge on kinetochores in some cell types since the negatively charged free ends of the adjacent kinetochore microtubule bundle will induce a positive charge on kinetochores. A calculation of the magnitude of the induced charge on kinetochores and centrosome matrices that is consistent with experimental values and is in accord with experimental observations of the magnitude of the poleward force on chromosomes during mitosis will now be carried out.

4.2   Induced charge on centrosomes and kinetochores

As mentioned above, although it is probable that kinetochores and centrosomes carry a net charge, assumptions of net charge may not be necesssary since charge at the ends of microtubules can induce charge on these structures. Thus net positive charge on the minus ends of microtubules in the polar, kinetochore, and astral microtubule bundles at a centrosome will induce negative charge on the centrosome matrix. Negative charge at the free ends of microtubules near kinetochores in a kinetochore microtubule bundle will induce positive charge on a kinetochore. In some cell types, induced charge at either a centrosome matrix or kinetochore, or at both, may be responsible for poleward chromosome motions. Additionally, both permanent and induced charge contributions are subject to modification by cell-cycle pH_i variations. The magnitude of the induced charge and the conditions under which charge induction can occur will be discussed next.

A standard derivation [see for example, Frankl, 1986] from electrostatics shows that a point charge q at a perpendicular distance x from a planar boundary between two dielectric materials of permittivity ε_1 and ε_2 will induce a polarization charge density $\sigma(x, s)$ (C/m^2) at the interface given by

$$\sigma(x, s) = \frac{q\,x}{2\pi k_1 \left(s^2 + x^2\right)^{3/2}} \left(\frac{\varepsilon_1 - \varepsilon_2}{\varepsilon_1 + \varepsilon_2}\right), \qquad (4.1)$$

where q is embedded in dielectric medium 1 of dielectric constant k_1 and σ is the induced charge per unit area on the interface at a perpendicular distance s from a line connecting q and its image charge q' in dielectric medium 2 as depicted in Figure 4.1.

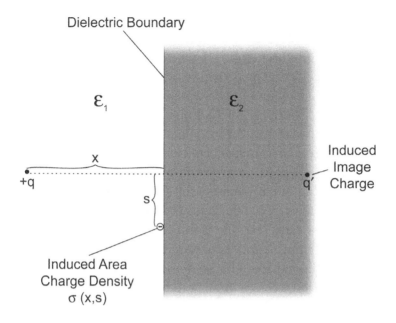

Dielectric Boundary

ε_1 ε_2

x

+q   q′

Induced Image Charge

s

Induced Area Charge Density σ (x,s)

Fig. 4.1. Geometry of induced electrostatic charge at a planar interface between the cytosol with permittivity ε_1 and a centrosome matrix or kinetochore with permittivity ε_2.

In the context of the present work, ε_1 is the permittivity of the cytosol (essentially cytoplasmic water) at a protofilament free end where a charge q is located and ε_2 is the permittivity of the dielectric medium (medium 2) within which the image charge is induced, a centrosome matrix or a kinetochore. As discussed in the next sections, a planar geometry is assumed for the cytosol-centrosome matrix or cytosol-kinetochore interfaces because a centrosome or kinetochore is much larger than the diameter of a protofilament free end on which the charge q is located.

Microtubule polymerization occurring at protofilament free ends with distances of 8 to 11 nm from either a centrosome matrix or a kinetochore could add 8 nm electric dipolar tubulin dimers, resulting in protofilament free ends at distances of 0 to 3 nm from the interface between the cytosol and a centrosome matrix or kinetochore. As discussed in Chapter 3, the 0–3 range of distances is significant for the present calculation because 1 nm may be taken as the thickness of the layered water adsorbed to each charged surface [Pauling, 1945; Pollack, 2001]; and as charged surfaces approach within 3 nm, counterion screening would be largely eliminated in the space between charged protofilament free ends and a charged centrosome matrix, as well as between charged protofilament free ends and a charged kinetochore. As also noted in Chapter 3, counterion (Debye) shielding between charged surfaces must be taken into account for surface separations greater than 3 nm; however, the following calculations will show that the induced charge density and the resulting poleward electrostatic force are greatly reduced at this distance even without counterion screening.

The magnitude of the induced charge density on a centrosome matrix or kinetochore due to a microtubule with its nearest protofilament free ends at distances of 1, 2, and 3 nm from a centrosome matrix or kinetochore will now be calculated. From (4.1), $\sigma(x, 0)$ ($=\sigma(x)$) at the interface between the cytosol and a point directly adjacent to the charge at a protofilament free end is

$$\sigma(x) = \frac{q}{2\pi k_1 \, x^2} \left(\frac{\varepsilon_1 - \varepsilon_2}{\varepsilon_1 + \varepsilon_2} \right) , \qquad (4.2)$$

where ε_1 ($= k_1 \varepsilon_0$) is the permittivity of layered cellular water at a protofilament free end and ε_2 ($= k_2 \varepsilon_0$) is the permittivity of a centrosome matrix or kinetochore.

As discussed previously, it is well established in electrochemistry that the permittivity of vicinal water layers outside a charged surface are much smaller than the bulk phase value. The dielectric constants $k_1(x)$ of water as a function of distance from a charged surface were measured by atomic force microscopy, and at distances of 1, 2, 3, and 4 nm were determined to have values of 9, 21, 40, and 60, respectively, increasing to close to the bulk value at a distance of 25 nm [Teschke *et al.*, 2001].

As discussed in Chapter 3, the value of the dielectric constant k_2 for a centrosome matrix or kinetochore has not been established and a cytosol-saturated centrosome matrix or kinetochore would be expected to have a dielectric constant that is roughly midway between their dry values and cytoplasmic water. From these considerations, the conservative value $k_2 = 30$ was chosen. It is necessary that k_2 be greater than the values of $k_1(x)$ for charged surface separations of 3 nm or less in order that the net induced charge be opposite in sign to the charge on the protofilament free ends ($k_1(x) < k_2$ in the above equations). It is easily verified from the experimental data given above that the interpolated values of $k_1(x)$ for separations of up to 5 nm between charged surfaces are less than the experimental values given above for one surface, and less than 30, so this will be the case here.

Calculation will show that protofilament free ends at distances greater than 3 nm from a centrosome matrix or kinetochore would be weakly interacting due to the rapid decrease of the induced charge. It is important to notice that the electrostatic force is repulsive at distances where $k_1(x) > k_2$; however, as we will see, electrostatic interaction energies are small at such distances as a result of the rapid decrease of the induced charge with increasing distance. The onset of counterion screening beyond 3 nm would decrease the interaction energy even further, allowing thermal energy to dominate at such distances and beyond. Tubulin polymerization at distances of 8 to 11 nm will therefore not be influenced by these electric fields.

For a kinetochore microtubule with the closest protofilament free end at a distance of 1 nm and a positive charge q of magnitude one electron charge at each protofilament free end, the maximum charge density $\sigma(x)$ induced on a centrosome matrix from positively charged protofilament free ends at x distances, irrespective of protofilament curling, of 1, 2, and 3 nm is found from (4.2) to be –2.3, –0.2, and –0.02 mC/m^2 respectively, for a total induced charge density σ of –2.5 mC/m^2. Similarly, for microtubules with the closest protofilament free end at distances of 2 or 3 nm, total induced charge densities are –0.22 and –0.02 mC/ m^2 respectively. A kinetochore microtubule with its closest protofilament free end at 4 nm will induce a negligibly small positive charge density, and even less if counterion screening is included.

It is easily verified from (4.1) that the induced charge density $\sigma(x, s)$ calculated for the most contributing x value of 1 nm falls off to 3 % of the maximum value at $s = 0$ for an s distance of 3 nm and to 17 % of the maximum for an s distance of 3 nm if $x = 2$ nm. Thus, since the radius of a protofilament can be taken as 2.5 nm, most of the image charge is induced over an area "in the shadow" of the cross sectional area of an approaching protofilament free end for the most contributing x distances. From experimental observations of the spacing between kinetochore microtubules, this implies that a growing kinetochore microtubule interacts primarily with the image charge that it induces locally on a kinetochore or centrosome matrix, and the induced charge from other kinetochore microtubules in the approaching kinetochore microtubule bundle is relatively small adjacent to that kinetochore microtubule.

4.3 Electrostatic microtubule disassembly force at cell poles

Observations on post-attachment chromosome movements, including the motive force at spindle poles, are explained within the context of induced charge electrostatics as follows. Microtubules are known to be in a constant condition of dynamic instability near the balanced state. From Chapter 1 and the discussion in Section 4.1, the net charge on protofilament free ends proximal to a centrosome matrix is positive. A γ-tubulin molecule at the surface of the centrosome matrix takes the form of a ring from which a kinetochore microtubule seems to emerge [Alberts *et al.*, 1994c]. This geometry could allow the electric field of the induced negative charge on γ-tubulin rings to draw the positively charged ends of kinetochore microtubules through the centrosome matrix, with the electric field gradient destabilizing the microtubules as they approach and pass through the charge distribution.

In Section 3.2 it was emphasized that, within the context of electrostatics, microtubules need not necessarily penetrate through the rings; rather the γ-tubulin rings embedded in the fibrous matrix comprise a firmly anchored negative charge distribution near the surface of the centrosome through which microtubules pass, disassembling in the process, as depicted in Figure 4.2, repeated here from Figure 3.1 for convenience.

As mentioned in the introduction, observations on a number of cell types have shown that disassembly of kinetochore microtubules at spindle poles accompanies chromosome poleward movement. Accordingly within the context of induced electrostatic charge, force generation at a spindle pole for prometaphase post-attachment, metaphase, and anaphase-A chromosome poleward movement can be attributed to an electrostatic attraction between the charged free ends of subsequently disassembling kinetochore microtubules and an induced opposite charge on the centrosome matrix [Gagliardi, 2008].

Since the outer diameter of a centrosome matrix is considerably larger than the diameter of a protofilament, we may model it as a large, approximately planar, slab. Integrating (4.1) [Frankl, 1986] to obtain the force between the charge q on a protofilament free end and the induced image charge on a centrosome matrix or kinetochore at a distance x from either surface, one finds that

$$F = \frac{1}{4\pi\varepsilon_1} \frac{(\varepsilon_1 - \varepsilon_2)}{(\varepsilon_1 + \varepsilon_2)} \frac{q^2}{4\,x^2} \, . \tag{4.3}$$

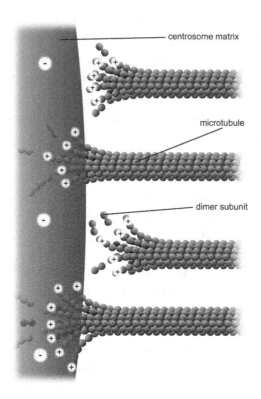

Fig. 4.2. Nanoscale electrostatic disassembly force at a centrosome. A poleward force results from an electrostatic attraction between positively charged microtubule free ends and induced negative charge on a centrosome matrix. Only disassembling microtubules are depicted; assembling microtubules could also be momentarily attracted to a centrosome.

From the geometry of microtubules outlined in Chapter 1, a kinetochore microtubule with its closest protofilament free end at a distance of 1 nm could also have protofilament free ends at the next closest distances of 2 nm and 3 nm. In accord with the discussion in Chapter 3, a conservative calculation of the maximum force exerted by a centrosome matrix on a kinetochore microtubule may be carried out with (4.3) by assuming a kinetochore microtubule with protofilament free ends at the closest distances of 1, 2, and 3 nm, irrespective of protofilament curling. Summing the contributions over x and the experimental values of $k_1(x)$, results in

$$F = -5.2\,n^2\ (\text{pN/MT})\,, \qquad (4.4)$$

where $q = ne$ has been substituted, with e equal to the charge on an electron and n the number of electron charges at or near a protofilament free end. Comparing this with the experimental range 1–74 pN/MT, we have that $n = 0.44 - 3.8$, well within the experimental range [Brown and Tuszyński, 1997; Tuszyński et al., 1998; Tuszyński et al., 1995; Stracke et al., 2002].

There does not appear to be consensus on a model for the generation of the force associated with microtubule flux at cell poles. Experimental observations regarding microtubule flux at poles are explained consistently within the context of the present work by the same nanoscale electrostatic microtubule disassembly force as that operating at kinetochores.

4.4   Electrostatic microtubule disassembly force at kinetochores

A calculation of the maximum poleward force on chromosomes has been carried out with negative charge at protofilament free ends of kinetochore microtubules interacting with equal positive charge at protofilament free ends on very short microtubule stubs embedded in a kinetochore [Gagliardi, 2002a]. This calculation was primarily intended to demonstrate that electrostatic force could reproduce the experimental results. A computer simulation for anaphase-A chromosome motion incorporating the geometry of microtubules along with a numerical integration of Newton's second law with typical values of chromosome mass [Lewin, 1978] and cytosol viscosity [Alexander and Rieder,

1991] indicates that electrostatic force is robust enough to sustain anaphase-A motion within a wide range of microtubule disassembly modes [Gagliardi, 2002a]. The simulation also confirms that the experimentally observed anaphase-A chromosome speeds of a few micrometers per minute are determined almost exclusively by the disassembly rate of microtubules over a wide range of disassembly modes and charge values.

As indicated above, based on electron micrographs, it is generally accepted that kinetochore microtubules penetrate the outer plates of kinetochores. It is also often assumed that this kinetochore microtubule association is the locus of force generation between kinetochores and microtubules. Consequently, not much attention has been given to the possibility that kinetochore microtubules may be generating force in non-contact interactions such as those arising from electrostatics. Force generation by electrostatic non-contact interactions, primarily over distances of a few nanometers – as discussed above – could cause other, previously force-generating, kinetochore microtubules in the bundle to penetrate into the open structure of the kinetochore, giving the illusion of contact force generation. This situation would also apply at a centrosome matrix.

We now proceed to calculate the poleward force due to non-penetrating microtubules at kinetochores. Since kinetochore plate diameters are large compared to the diameters of protofilaments, we may model the kinetochore–protofilament interaction by assuming a large approximately planar slab for the kinetochore as depicted in Figure 4.3, repeated here for convenience from Figure 3.2. Because of the similarity in geometry, a calculation of the maximum force per microtubule at a kinetochore for non-penetrating microtubules will yield essentially the same result as the calculation at a centrosome matrix. The dry dielectric constant k_2 of a kinetochore is likely to differ from the value for the centrosome matrix. However, as discussed for the centrosome matrix, given the open structure of a kinetochore with cytoplasmic water intrusion, the dielectric constant would be expected to lie between that of water and the dry kinetochore.

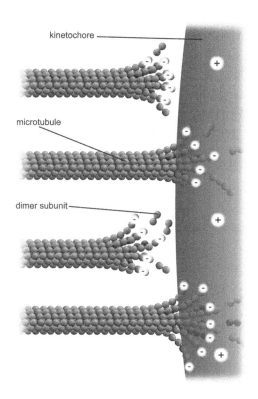

Fig. 4.3. Nanoscale electrostatic disassembly force at a charged kinetochore. A poleward force results from an electrostatic attraction between negatively charged microtubule free ends and induced positive charge on the kinetochore. Only disassembling microtubules are depicted; assembling microtubules can also momentarily generate poleward force.

As with the centrosome matrix, (1) the value for cytoplasmic water will dominate, and (2) the force calculation is relatively insensitive to the precise dry value, except that k_2 must be greater than k_1 for the electrostatic force due to induced charge to be attractive. It will therefore be convenient to assume the same conservative estimate of 30 that was used for the centrosome matrix.

As at a centrosome matrix, an identical force calculation with (4.3) yields values of n between 0.44 and 3.8, well within the experimental range. The calculation given here may be regarded as an *ab initio* calculation of the maximum force per microtubule; however, as previously indicated, these model calculations are primarily intended to demonstrate that electrostatic interactions are able to produce a force within the experimental range.

It was shown above that the magnitude of the induced charge is a rapidly changing function of the protofilament free end distance to a centrosome matrix or kinetochore for separations between 0 and 3 nm. It therefore follows that the field gradient over a short range of distances both inside and outside the pole-facing outer kinetochore plate could destabilize the protofilaments in the vicinity of kinetochores. As in the case for protofilaments approaching and penetrating a charged centrosome matrix, this could cause kinetochore microtubules to disassemble near – both inside and outside – the pole-facing outer kinetochore plate as poleward force is generated, in agreement with experimental observation.

4.5 Summary

Calculations of force magnitudes for both penetrating and non-penetrating microtubules within critical separations support the possibility that, given the known net charge at the free ends of microtubules, induced charge on kinetochores and centrosome matrices could be responsible for poleward chromosome motion. The calculated maximum force per microtubule falls within the experimentally measured range.

In agreement with experiment [Nicklas, 1988], the calculations given in this chapter satisfy the requirement that the maximum tension force per microtubule be the same for all mitotic post-attachment chromosome movements. The approximate equality in the calculations for electrostatic force generation by non-penetrating kinetochore microtubules at poles and kinetochores demonstrated here for induced charge on kinetochores and centrosome matrices is to be expected of any credible force generating mechanism for poleward chromosome motions.

Models for chromosome movements must also address the efficiency with which dynamic spindle microtubules maintain coupling to kintochores and poles throughout the various movements during mitosis. Electrostatic force at nanometer distances between known charge at the free ends of a kinetochore microtubule and induced charge on a kinetochore or centrosome matrix has this property.

Consistent with observation, spindle microtubules depolymerize while generating force at kinetochores and poles. Given the electrostatic nature of tubulin microtubule subunits, this can be understood within the context of electrostatic interactions in terms of the large electric field (and therefore force) gradients in the critical distances within vicinal cytosol outside and across the boundaries of kinetochores and centrosome matrices.

Chapter 5

Electrostatics in Mitotic Chromosome Motions

5.1 Introduction

Poleward and antipoleward chromosome movements occur intermittently during prometaphase and metaphase. Antipoleward motions dominate during the *congressional* movement of chromosomes to the cell equator. Poleward motion of chromosomes dominates during anaphase-A. The apparent complexity of these motions has challenged scientific explanation for over a hundred years. While a number of models have been advanced for individual motions, such as anaphase-A, it can be argued that the various motions are related, and that a simple theme or principle that unifies mitotic motions should be sought. It is proposed in this work that this can be accomplished by attributing the cause for all post-attachment chromosome motions to microtubule dynamics in combination with nanoscale electrostatics.

In present terminology *metaphase* denotes the relatively brief period during which chromosomes are lined up at the center of the cell (the *equator*) and are fully attached to both poles by the microtubules of the spindle, whereas *prometaphase* is used to encompass a much wider time period during which most of the complex motions in this stage of mitosis occur. Two events that are of major significance during prometaphase are (1) the *capture* and *attachment* of chromatid pairs by microtubules, and (2) chromosome movement to, and alignment at, the cell equator. The latter is comprised of several distinguishable motions.

Regarding the first event, experiments [Rieder and Alexander, 1990] have shown that during prometaphase each pair of sister chromatids attaches by a kinetochore to the outside walls of a single microtubule, resulting in a rapid microtubule sidewall sliding movement toward a pole. This motion is postulated to be driven by molecular motors. A molecular motor-powered microtubule sidewall sliding model for this prometaphase movement would appear likely. In particular, the speed (20–50 μm per minute) [Grancell and Sorger, 1998] of kinetochores along microtubule walls is consistent with known molecular motor behavior.

Consequently, it is likely that a molecular motor model for the microtubule sidewall motion associated with chromosome *capture* is supported by the experimental observations. However, it is probable that post-attachment prometaphase, metaphase and anaphase-A chromosome motions can be understood in terms of a cell-cycle dependent increase in the dominance of nanoscale electrostatic microtubule disassembly forces.

This increase through metaphase may be due to a rising microtubule disassembly/assembly probability ratio caused by a steadily decreasing pH_i throughout mitosis. As will be discussed, any further possible decrease in pH_i beyond metaphase may not be the major determining factor in chromosome dynamics.

As discussed in Chapters 3 and 4, poleward motions could be due to charged plus and minus ends of kinetochore microtubules interacting with permanent or induced charge on kinetochores and centrosomes, respectively. The calculations in Chapter 4 show that it may not be necessary to assume permanent charge on kinetochores or centrosomes since charge induced by known charge at both ends of microtubules is sufficient. As mentioned previously, it is possible that in some cell types, induced charge at either a centrosome matrix or kinetochore, or at both, may be responsible for poleward chromosome motion. Both permanent and induced charge density magnitudes are subject to pH_i variations.

Antipoleward nanoscale microtubule assembly forces will now be considered. As a result of the sliding capture motion described above, the approach to the poles will result in the movement of kinetochores to within critical distances from the ends of other *astral* microtubules emanating from the closer pole. As discussed in previous chapters, counterion screening is not fully operative over small separation distances up to approximately 3 nm. The resulting proximity – in conjunction with (1) an electrostatic attraction between positively charged kinetochores and the negatively charged ends of astral microtubules, and (2) an electrostatic repulsion between negatively charged chromosome arms in the chromatid pair and other negatively charged astral microtubule ends – could be critical in the orientation and attachment of kinetochores to the free plus ends of microtubules [Gagliardi, 2002a].

Following this *monovalent* attachment to one pole, chromosomes are observed to move at considerably slower speeds, a few μm per minute, in subsequent motions throughout prometaphase [Grancell and Sorger, 1998]. In particular, a period of slow motions toward and away from a pole will ensue, until close proximity of the negatively charged end of a microtubule from the opposite pole with the other (*sister*) kinetochore in the chromatid pair results in an attachment to both poles (a *bivalent* attachment). Attachments of additional microtubules from both poles will follow. (There may have been additional attachments to the first pole before any attachment to the second.)

After the sister kinetochore becomes attached to microtubules from the opposite pole, the chromosomes perform a slow (1–

$2\,\mu$m per minute) congressional motion to the spindle equator, resulting in the well-known metaphase alignment of chromatid pairs. In addition to the mechanism facilitating attachment just discussed, all of the above mentioned experimentally observed post-attachment poleward and antipoleward prometaphase motions, as well as oscillatory metaphase motion, can be understood in terms of electrostatic interactions coupled with microtubule dynamics.

Chromosome motion during anaphase has two major components, designated as anaphase-A and anaphase-B. Anaphase-A is concerned with the poleward motion of chromosomes, accompanied by the shortening of kinetochore microtubules at kinetochores and/or spindle poles. The second component, referred to as anaphase-B, involves the separation of the poles. Both components contribute to the increased separation of chromosomes during mitosis. Anaphase-A and anaphase-B will be addressed in terms of electrostatics later in this chapter.

5.2 Antipoleward nanoscale electrostatic assembly force

Since chromosome arms are negatively charged, following chromosome attachment they will be repelled from the negatively charged free ends of the shorter astral microtubules in the polar region. As mentioned above and discussed in Chapter 1, this force will be effective over the critical distances allowed by modified Debye screening. As microtubules assemble, and chromosomes move farther from the poles, there will be a filling in of dipolar subunits in the gaps between astral microtubule free ends and chromosome arms.

Polymerization will take place in the gaps opened up by electrostatic repulsion between negatively charged microtubule free plus ends and negatively charged chromosome arms as they drift farther from the poles, and chromosomes will be continuously repelled from the poles. This mechanism may account for the antipoleward *astral exclusion force*, or *polar wind*, the precise nature of which has been sought since it was first observed [Rieder *et al.*,1986]. The interaction between astral microtubules and chromosome arms is depicted in Figure 5.1.

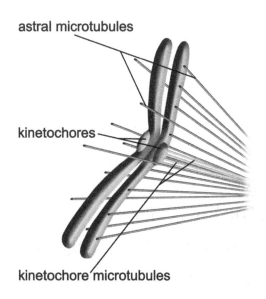

astral microtubules

kinetochores

kinetochore microtubules

Fig. 5.1. Antipoleward nanoscale electrostatic force between microtubules and chromosome arms. An antipoleward force results from electrostatic repulsion between negatively charged plus ends of microtubules and negatively charged chromosome arms.

Short range repulsive entropic forces associated with growing microtubules would complement the electrostatic repulsive interaction at small microtubule–chromosome arm separations, adding to the total astral exclusion force. Although the complex geometry precludes a theoretical calculation of the magnitude of these forces, a model calculation of the repulsive force between two like charged parallel surfaces with an electrolyte in between shows that entropic forces must be included for separations of less than 2 nm; at greater separations electrostatic theory fits the data well [Israelachvili, 1991; Cowley, 1978].

As a chromatid pair moves farther from a pole, electrostatic repulsive force between the negatively charged free ends of astral microtubules and chromosomes will decrease as the microtubules fan radially outward (Figure 5.2).

At a surface defined by the microtubule ends, the charge density and therefore the force, will decrease according to an "inverse square law," as we can see from the following. Given that the repulsive force on a chromosome arm depends on the total number N of negatively charged microtubule free ends from which it is repelled, we have $F \sim Nq$, where q is the charge at the end of a microtubule.

For N microtubules fanning radially outward from a pole, the total charge Nq is distributed over an area that increases as the distance r from the pole squared (r^2), and the effective charge per unit area at a surface defined by the microtubule ends decreases as the inverse of the distance squared $(1/r^2)$. This results in an electrostatic antipoleward force on chromosome arms that decreases with an *inverse square* $(1/r^2)$ dependence on the polar distance.

The falloff is expected to be even more pronounced than inverse square would predict because of the decreased number of microtubule free ends at greater polar distances, as shown schematically in Figure 5.2, resulting in an even stronger falloff of antipoleward force with distance from cell poles. To save writing, this stronger than inverse square falloff will be characterized as *robust inverse square*.

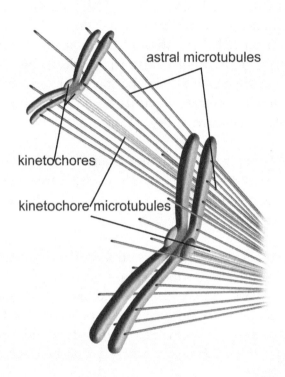

Fig. 5.2. Antipoleward inverse square repulsive force. Two chromatid pairs at differing polar distances are depicted showing the robust inverse square dependence of the nanoscale antipoleward force.

5.3 Prometaphase and metaphase chromosome motions

Microtubule polymerization or depolymerization, in combination with a repulsive electrostatic antipoleward astral exclusion force and an attractive electrostatic poleward directed force acting at kinetochores and spindle poles, is sufficient to account for the observed motion of monovalently attached chromosomes in terms of electrostatics. Because of statistical fluctuations both in the number of disassembling kinetochore microtubules interacting with kinetochores as well as centrosomes, and in the number of assembling astral microtubules responsible for the antipoleward astral exclusion force, the interaction of these opposing forces could result in a "tug of war," consistent with the experimentally observed series of movements toward and away from a pole for a monovalently attached chromatid pair.

Microtubule assembly at kinetochores and poles is also possible; however, because the necessary (see below) inverse square dependence of the antipoleward force cannot be derived from microtubule assembly at kinetochores or spindle poles, it is assumed in this work that assembly at either location is in passive stochastic response to assembly at chromosome arms, or to tension caused by poleward force on a sister kinetochore.

After a bivalent attachment has been established, the attractive force to the far (*distal*) pole will be in opposition to the attractive force to the near (*proximal*) pole. Because of the robust inverse square astral exclusion force, the greater repulsion from the proximal pole, along with a growing number of kinetochore attachments to microtubules from the distal pole tending to equalize poleward disassembly forces, a relatively sustained congressional motion away from the proximal pole would result, as observed experimentally.

As a chromatid pair moves farther from the proximal pole, there will be a growing number of attachments to both poles. Following approximately equal numbers of attachments to both poles, and comparable distances of chromatid pairs from the two poles, the forces exerted by both sets of poleward attractive disassembly and antipoleward repulsive assembly forces will approach equality. Thus, as a chromatid pair congresses to the midcell region, the number of attachments to both poles will tend to be the same, as will the number of microtubules interacting with chromosome arms, and equilibrium of poleward directed forces and antipoleward astral exclusion forces will be approached. Without specifying their exact nature, balanced pairs of attractive and repulsive forces have previously been postulated for the metaphase alignment of chromatid pairs [Alberts *et al.*, 1994d].

An explanation of experimentally observed metaphase oscillations about the cell equator just prior to anaphase-A provides another example of the predictability and minimal assumptions nature of the present approach. In agreement with experiment [Hays and Salmon, 1990], an electrostatic force model predicts that the poleward force on a chromosome from kinetochore microtubule disassembly at kinetochores or poles depends on the total number of kinetochore microtubules. At the metaphase "plate," the bivalent attachment of chromatid pairs ensures that the poleward-directed electrostatic disassembly force on one chromatid at a given moment could be greater than that at the sister chromatid's kinetochore attached to the opposite pole. An imbalance of these poleward forces results from statistical fluctuations in the number of force generating microtubules at kinetochores as well as at poles.

This situation, coupled with similar fluctuations in the number of astral microtubules responsible for the antipoleward astral exclusion force, can result in a momentary motion toward a pole in the direction of the instantaneous net electrostatic force. However, because of the robust inverse square dependence of the repulsive astral exclusion force and the approximate equality of poleward-directed microtubule disassembly forces for chromatid pairs in the midcell region, the greater force of repulsion from the proximal pole will eventually reverse the direction of motion, resulting in stable equilibrium midcell metaphase oscillations, as observed experimentally.

As cited in Chapter 1, experiments have shown that the intracellular pH (pH_i) of many cell types rises to a maximum at the onset of mitosis, subsequently falling steadily through mitosis. Although it is experimentally difficult to resolve the exact starting time for the beginning of the decrease in pH_i during the cell cycle, it appears to decrease 0.3 to 0.5 pH units from the typical peak values of 7.3 to 7.5 measured earlier during prophase. The further decrease in pH_i through metaphase [Steinhardt and Morisawa, 1982] would result in increased instability of the microtubules comprising the spindle fibers. Previously, I noted that *in vivo* experiments have shown that microtubule stability is related to pH_i, with a more basic pH favoring microtubule assembly. An increased probability for microtubule depolymerization, as compared to the prophase predominance of microtubule assembly, is consistent with alternating poleward and antipoleward motions of monovalently attached chromosomes during prometaphase.

As discussed above, after a bivalent attachment, poleward forces toward both poles acting in conjunction with robust inverse square antipoleward forces could account for chromosome congression. The relative complexity of microtubule disassembly force generation at kinetochores and poles coupled with robust inverse square forces from microtubule assembly at chromosome arms precludes an unequivocal conclusion regarding a possible continuing increase in the microtubule disassembly to assembly probability ratio during chromosome congression. However, midcell metaphase oscillations are direct experimental evidence for a continuing increase in the disassembly/assembly probability ratio resulting in parity for microtubule assembly and disassembly probabilities.

At late metaphase, before anaphase-A, experiments reveal that the poleward motions of sister kinetochores stretch the intervening centromeric chromatin, producing high kinetochore tensions. It is reasonable to attribute these high tensions to a continuing disassembly to assembly probability ratio increase caused by a further lowering of pH_i. The resulting attendant increase in poleward electrostatic disassembly force would lead to increased tension. A lower pH_i will also increase the expression of positive charge on sister kinetochores, further increasing the tension due to their increased mutual repulsion. At these high tensions, coupled microtubule plus ends often switch from a depolymerization state to a polymerization state of dynamic instability. This may be explained by kinetochore microtubule free ends passively taking up the slack by polymerization to sustain attachment and resist further centromeric chromatin stretching. This is known as the "slip-clutch mechanism" [Maiato *et al.*, 2004].

This mechanism is addressed within the context of the present work as follows. Microtubule assembly at a kinetochore or pole is regarded here as operating in passive response to (1) the robust inverse square electrostatic antipoleward force acting between the plus ends of astral microtubules and chromosome arms and/or (2) an electrostatic microtubule disassembly force at a sister kinetochore or at poles. At the highest tensions, non-contact electrostatic forces acting over a range of protofilament gap distances would be effective in helping to maintain coupling while other larger protofilament gaps in the same or other microtubules are passively filled in. This process would continue with new gaps and new opportunities for fill-in. In addition, the repulsive robust inverse square electrostatic assembly force acting at the sister chromatid's arms will provide a positive feedback mechanism to resist detachment. This explanation of the slip-clutch mechanism follows as a direct consequence of the present approach to chromosome motility. The slip-clutch mechanism does not appear to be addressed by models for chromosome motions in the current literature.

In summary, regarding post-attachment chromosome movements through metaphase, it would seem reasonable to ascribe the increasing microtubule dissassembly/assembly probability ratio, with attendant changes in microtubule dynamics and mitotic chromosome motions, to an experimentally observed steadily decreasing pH_i. We may then envision the decrease in pH_i from a peak at prophase favoring microtubule assembly, declining through prometaphase, and continuing to decline through metaphase when parity between microtubule assembly and disassembly leads to midcell chromatid pair oscillations, culminating in increased disassembly tension near anaphase, as the cell's master clock controlling microtubule dynamics, and consequently the events of mitosis. One might also be tempted to attribute the more complete dominance of microtubule disassembly – with an accompanying predominance of poleward electrostatic disassembly forces – during anaphase-A to a further continuation of a decreasing intracellular pH. However, as we now discuss, any additional possible decreases in pH_i during anaphase-A may not be the major determinant in anaphase-A motion.

5.4   Anaphase-A chromosome motion

As discussed above, antipoleward electrostatic forces compete stochastically with poleward electrostatic forces during prometaphase and metaphase. For example, after a bivalent attachment is established, the action of poleward-directed forces from both poles, in conjunction with the robust inverse square nature of the antipoleward force, is sufficient for congressional motion to the cell equator followed by midcell metaphase chromosome oscillations. These oscillations were seen as resulting from inverse square antipoleward electrostatic assembly forces at chromosome arms acting in conjunction with approximately balanced poleward electrostatic disassembly forces. This balance is changed by subsequent events as will now be discussed.

In a number of cell types, intracellular calcium releases show a temporal correlation with the onset of anaphase-A. Experimental studies have shown that intracellular $[Ca^{2+}]$ increases are associated with anaphase-A chromosome movement [Hepler and Callaham, 1987; Hepler, 1989; Zhang et al., 1990]. It is well known that increased $[Ca^{2+}]$ facilitates the depolymerization of spindle microtubules both in vitro [Salmon and Segall, 1980] and in vivo [Kiehart, 1981]. These experimental observations have a direct interpretation within the framework of electrostatic force generating mechanisms.

With the observed increase in $[Ca^{2+}]$ at the onset of anaphase-A and the resultant further increase in the instability of microtubules, the disassembly to assembly probability ratio will be increased even more. After chromatid separation heralds the beginning of anaphase-A, the increased disassembly/assembly probability ratio allows nanoscale electrostatic microtubule disassembly forces at kinetochores and poles to dominate, enabling the dynamics for anaphase-A chromosome movement. At the highest net disassembly rates, there will be even less opportunity for kinetochore microtubule reassembly since steadily advancing kinetochores can more frequently shorten the gaps in regions previously occupied by dimer subunits, further reducing possible reassembly. Similarly, poleward microtubule motion (*flux*) due to electrostatic attraction of microtubules to centrosome matrices could minimize subsequent reassembly there.

Given that kinetochores exhibit a net positive charge and will manifest an even greater positive charge at lower pH_i levels in late metaphase, the resulting increased expression of charge on kinetochores, coupled with their close proximity and the inverse square nature of the Coulomb electrostatic force – acting in combination with increased electrostatic disassembly force tension from kinetochores and poles – could supply sufficient force to initiate chromatid separation. The observed increased kinetochore tension close to the metaphase-anaphase transition discussed above is consistent with this scenario. As of this writing, scientific consensus is lacking on a model explaining the timing and dynamics of chromatid separation, which would appear to be a natural deduction here.

Studies have shown that changes in $[Ca^{2+}]$ can modulate the speed of chromosome motion [Zhang *et al.*, 1990; Cande, 1981]. Significantly, there appears to be an optimum concentration for maximizing the speed of chromosome motions during anaphase-A. If $[Ca^{2+}]$ is increased to a micromolar level, anaphase-A chromosome motion is increased two-fold above the control rate;

however, if the concentration is further increased beyond a few micromolar, the chromosomes will slow down, and possibly stop [Zhang *et al.*, 1990]. It has long been recognized that one way elevated [Ca^{2+}] could increase the speed of chromosome motion during anaphase-A is by facilitating microtubule depolymerization [Weisenberg, 1972; Salmon and Segall, 1980; Kiehart, 1981; Cande, 1981; Olmsted and Borisy, 1975], and it is commonly believed that the breakdown of microtubules, if not the motor for chromosome motion, is at least the rate-determining step [Nicklas, 1975; Nicklas, 1987; Salmon, 1975; Salmon, 1989]. However, the slowing or stopping of chromosome motion associated with moderate increases beyond an optimum [Ca^{2+}] is more difficult to interpret since the microtubule network of the spindle is not compromised to the extent that anaphase-A chromosome motion could be slowed or stopped; this would require considerably higher concentrations [Wolfe, 1993c; Zhang *et al.*, 1990].

Experimental observation that an increase in calcium levels beyond micromolar levels results in a slowing or stopping of anaphase-A motion is a direct consequence of an electrostatic motor for mitotic chromosome motions. Higher concentrations of doubly charged calcium ions would screen the negative charge at the free ends of disassembling kinetochore microtubules and at the centrosome matrix, shutting down the poleward-directed nanoscale electrostatic disassembly force. Since this happens at concentrations that do not compromise the spindle's microtubule network, it is reasonable to interpret these results as experimentally consistent with an electrostatic motor for poleward chromosome motions.

An experimental test of nonspecific divalent cation effects on anaphase-A chromosome motion in which Mg^{2+} was substituted for Ca^{2+} [Zhang *et al.*, 1990] does not offer a definitive test for the possibility of negative charge cancellation by positive ions. The reason is that the positive charge of Mg^{2+} is shielded much more effectively by water than is the case for Ca^{2+}. This is shown by high frequency sound absorption studies of substitution rate constants for water molecules in the inner hydration shell of various ions which reveal that the inner hydration shell water substitution rate for Mg^{2+} is more than three orders of magnitude slower than that for Ca^{2+} [Diebler et al., 1969]. Thus, the slowing or stopping of anaphase-A chromosome motion accompanying free calcium concentration increases above the optimum concentration for maximum anaphase-A chromosome speed – but well below concentration levels that compromise the mitotic apparatus – is consistent with an electrostatic disassembly motor for poleward chromosome motions. This experimental result has not been addressed by any of the current models for anaphase-A motion.

5.5 Anaphase-B chromosome motion

The timing and dynamics of anaphase-B motion can also be explained within the context of nanoscale electrostatics. Experiments have revealed that polar microtubules overlap in the region of the central spindle in such a manner that the nearest neighbor of any microtubule in the overlap zone is likely to be a microtubule originating from the opposite pole [Pickett-Heaps, 1991]. Consistent with the predominance of kinetochore microtubule disassembly over assembly during anaphase-A, polar microtubules would also favor disassembly at this point in the cell cycle due to the Ca^{2+} release and low pH_i.

When the first subset of microtubules has shortened to an extent that the negatively charged free ends of polar microtubules eminating from opposite poles are within critical separations of each other, electrostatic repulsive forces between these ends can push the microtubules in this subset farther apart. These forces would continue to act until the free plus ends of microtubules in this subset are repelled to beyond critical separations. Such interacting microtubules will be continually replaced by other subsets whose ends are within critical separations, resulting in anaphase-B cellular elongation. The apparent rates of shortening vs. lengthening of polar microtubules may be different from kinetochore microtubules since, as indicated above, shortened gaps that result from sustained motions of kinetochores into regions previously occupied by tubulin dimers would reduce reassembly. A few interacting microtubules are represented schematically in a small two-dimensional section in Figure 5.3.

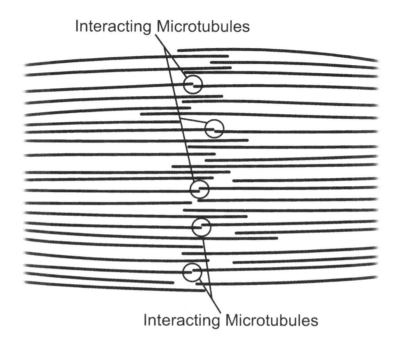

Fig. 5.3. A subset of interacting microtubules in a small central portion of an anaphase-B mitotic spindle. The free plus ends of interacting microtubules within a few nanometers are mutually repelling. Protofilament curling of disassembling microtubules is not shown on this scale.

Variable time delays between the onsets of anaphase-A and anaphase-B motions may be attributed to the time that it takes for the first subset of polar microtubule plus ends to go through a number of cycles of assembly and disassembly before shortening sufficiently from previously overlapping configurations for their free ends to interact as discussed above. Thus dynamic instability of polar microtubules – with net plus end disassembly, and net minus end assembly [Tippit *et al.*, 1978] at poles favored – could generate a steady repulsive force between like-charged free ends of interacting polar microtubules in the mitotic half-spindles. This would lead to a steadily decreasing amount of microtubule overlap with relatively constant half-spindle lengths, resulting in anaphase-B pole separation.

In a series of experiments utilizing labelling techniques [Masuda *et al.*, 1988], it was demonstrated that the minimum distance by which the poles separate corresponds to the amount by which microtubule overlap in the central spindle decreases, a finding that is consistent with the above explanation. These experiments also showed that the addition of ATP caused the poles of mitotic spindles *in vitro* to move apart, indicating that a molecular motor was possibly responsible for the motive force. The authors of this study concluded that the motor for anaphase-B is probably based on kinesin, and operates by a sliding mechanism between the microtubules.

However, in an experiment involving destruction of a portion of the central spindle by a laser microbeam [Aist *et al.*, 1991], the speed of anaphase-B motion increased by a factor of three, as if the central spindle retards the motion instead of causing it. This finding is difficult to explain if molecular motors pushing off between adjacent overlapping microtubules are responsible for the motion. Finally, experiments testing *polar tension* models for anaphase-B suggest that possible polar tension forces acting outside the spindle could be responsible for anaphase-B motion only in limited groups of organisms [Wolfe, 1993d].

The laser microbeam experiment may be understood within the present context since producing a gap – and the resulting local pool of highly concentrated tubulin dimer subunits available for polymerization – would temporarily significantly decrease the microtubule disassembly/assembly probability ratio to soon expose an even larger subset of interacting and potentially inter-

81

acting polar microtubules, with the consequent increase in total force and speed of separation.

5.6 Summary

Post-attachment chromosome motions during prometaphase and metaphase can be explained by statistical fluctuations in nano-scale electrostatic microtubule antipoleward assembly forces acting between microtubules and chromosome arms, combined with similar fluctuations in nanoscale electrostatic microtubule poleward disassembly forces acting at kinetochores and spindle poles. The different motions throughout prometaphase and metaphase may be understood as an increase in the microtubule disassembly to assembly probability ratio. It seems reasonable to assume that the shift from the dominance of microtubule growth during prophase, and to a lesser extent during prometaphase, to a parity between microtubule polymerization and depolymerization during metaphase chromosome oscillations could be attributed to the gradual downward pH_i shift during mitosis that is observed in many cell types.

Evidence for a continuing decrease in pH_i and an increasing microtubule disassembly to assembly probability ratio is also seen in increased kinetochore tension just prior to anaphase. This increased tension has a possible simple interpretation in terms of the greater magnitude of poleward electrostatic disassembly forces at sister kinetochores and poles relative to antipoleward assembly forces, as well as an increased mutual repulsion of sister kinetochores due to a greater expression of kinetochore positive charge, both due to decreased pH_i.

There does not appear to be consensus on a model for the generation of poleward force at cell poles. Experimental observations regarding the microtubule disassembly force at poles, with an associated microtubule flux, can be consistently explained in terms of the same nanoscale electrostatic force mechanism as that operating at kinetochores.

The factors: (1) an intracellular pH decrease continuing through mitosis, (2) an electrostatic component to microtubule disassembly/assembly probability ratios, (3) a nanoscale electrostatic microtubule disassembly force acting at kinetochores and centrosome matrices, and (4) a nanoscale electrostatic microtubule assembly force acting between negatively charged microtubule plus ends and like charged chromosome arms, all working in conjunction with microtubule dynamics, make it possible to explain the motive force – and timing – for chromosome motions throughout post-attachment prometaphase and metaphase.

The observed intracellular increase in $[Ca^{2+}]$ that occurs at the onset of anaphase-A further increases the probability for microtubule disassembly, effectively dwarfing antipoleward nanoscale microtubule assembly forces from both poles. The sudden dominance of poleward nanoscale microtubule disassembly forces added to a greater mutual repulsion of positively charged sister kinetochores, with both due to a decreased pH_i could be integral in the initial separation of sister chromatids. Once this separation is effected, anaphase-A motion would result from the predominance of electrostatic microtubule disassembly forces at kinetochores and poles.

Critical experimental results such as the slip-clutch mechanism and observations of calcium ion concentrations on anaphase-A chromosome movements are not addressed by models in the current literature. Both experimental observations are explainable within the context of the present approach to chromosome motility.

In addition to destabilizing kinetochore microtubules for anaphase-A motion, the free calcium release would also destabilize the plus ends of polar microtubules, favoring their depolymerization. Depending on the cell type, it can take some time for the free ends of sufficient numbers of polar microtubules from opposite poles to stochastically depolymerize to within critical separations between their respective like-charged plus ends. It is reasonable to expect that shortening rates for kinetochore microtubules would be greater than those for polar microtubules. This is because poleward advancing kinetochores will decrease microtubule polymerization opportunities in shortening gaps.

Anaphase-B motion can therefore be attributed to an electrostatic repulsion between adjacent negatively charged plus ends of constantly changing subsets of microtubules eminating from opposite poles disassembling from a geometry of previously overlapping polar microtubules while a constant half-spindle length is maintained due to minus end assembly.

Appendix A

Some results from electrostatics

The electric potential ϕ (potential energy per unit charge) of a point charge in the field of another point charge q_1 is given by $q_1/4\pi\varepsilon r$, where r is the distance between the point charges. If the two charges represent spherically symmetric charge distributions of radii a and b, then r (where $r > (a + b)$)is the distance between their cenxters. The electric potential energy U of q_2 in the field of q_1 is $q_1 q_2/4\pi\varepsilon r_{12}$, where r_{12} ($= r_{21}$) is either of the distances mentioned mentioned above. Because of the obvious symmetry of the situation, this is more properly described as the electric potential energy of the system of charges consisting of q_1 and q_2. Similarly, for three charges, q_1, q_2, and q_3, we have

$$U = q_1\,q_2/4\pi\varepsilon r_{12} + q_1\,q_3/4\pi\varepsilon r_{13} + q_2\,q_3/4\pi\varepsilon r_{23}\,. \qquad (A.1)$$

This equation may be written as

$$U = \frac{1}{2}\left[\frac{q_1\,q_2}{4\pi\varepsilon r_{12}} + \frac{q_2\,q_1}{4\pi\varepsilon r_{21}} + \frac{q_1\,q_3}{4\pi\varepsilon r_{13}} + \frac{q_3\,q_1}{4\pi\varepsilon r_{31}} + \frac{q_2\,q_3}{4\pi\varepsilon r_{23}} + \frac{q_3\,q_2}{4\pi\varepsilon r_{32}}\right]$$

Generalizing, the electric potential energy of a system of point charges is given by

$$U = \frac{1}{2}\sum_{i,j}\frac{q_i\,q_j}{4\pi\varepsilon r_{ij}}\,, \qquad (A.2)$$

where the sum is carried out over all possible charge pairs, and $i \neq j$. Equivalently, this may be written as

$$U = \frac{1}{2} \sum_{i=1}^{N} q_i \, \phi_i \,, \tag{A.3}$$

where ϕ_i is the electric potential due to all the other charges in the system at the point occupied by the charge q_i.

For continuum charge distributions, the differential charge dq replaces q_i, and ϕ replaces ϕ_i in the above equation. Thus, we have

$$U = \frac{1}{2} \int \rho \phi \, d\tau, \tag{A.4}$$

where ρ is the volume charge density, $d\tau$ is the volume element, $dq = \rho \, d\tau$, and the integration is over any volume containing all the charges in the system.

For a surface charge distribution with area charge density σ, we have $dq = \sigma \, dA$, and dA an element of area on the surface, this becomes

$$U = \frac{1}{2} \int \sigma \phi \, dA \,. \tag{A.5}$$

We now use this relation to derive an expression for the electrostatic self energy of a uniformly charged spherical shell.

For a uniformly charged spherical shell of radius R_0 and charge Q, we have the relationship $\phi = Q/4\pi\varepsilon r$, where $r \geq R_0$. At the surface, $r = R_0$, the electric potential equals $Q/4\pi\varepsilon R_0$, and is therefore constant over the spherical surface. We may then write for the electrostatic energy of the spherical shell

$$U = \frac{1}{2} \frac{Q}{4\pi\varepsilon R_0} \int \sigma \, dA = Q^2/8\pi\varepsilon R_0 \,, \tag{A.6}$$

since $\int \sigma \, dA = Q$.

Appendix B

Debye theory of counterion screening

The concentration c_i of oppositely charged ions around an ion or charged particle as a function of distance r from the charge will depend on the electrostatic potential energy $U(r)$ of the ions in the field of the central charge. This dependence is given by a classical Boltzmann factor $Ae^{-U(r)/kT}$, where k is Boltzmann's constant, T is the absolute temperature, and A is a constant. At a large distance, $r \to \infty$, from the charge, $U(r) \to 0$, and the concentrations of the ions in the solution approach their values in the bulk electrolyte, $c_i(\infty)$. Therefore, at a distance r from the ion or charged particle i, the concentration of oppositely charged ions (number of ions per cubic meter, m^{-3}) is

$$c_i(r) = Ae^{-U(r)/kT}, \qquad (B.1)$$

and $c_i(\infty) = Ae^0 = A$. Substituting this value for A in the previous equation, we we have

$$c_i(r) = c_i(\infty)e^{-U(r)/kT} \qquad (B.2)$$

From Appendix A, the electric potential energy $U(r)$ of an ion with charge q_i is given by $q_i\,\phi(r)$, where $\phi(r)$ is the electric potential (electric potential energy per unit charge). Since $q_i = z_i\,e$, with z_i the valence of the ion and e the magnitude of the electronic charge, we have

$$c_i(r) = c_i(\infty)\,e^{-z_i e\phi(r)/kT}. \qquad (B.3)$$

The electric charge density $\rho(r)$ at a distance r is therefore given by

$$\rho(r) = \sum_i z_i e c_i(r) = e \sum_i z_i c_i(\infty) e^{-z_i e \phi(r)/kT}. \qquad (B.4)$$

Note that the overall electrical neutrality from all the ions in solution dictates that $\rho(\infty) = \sum_i e z_i c_i(\infty) = 0$, since $\phi(r \rightarrow \infty) \rightarrow 0$. Therefore, we retrieve the expected overall electrical neutrality of the electrolyte far from the given charged particle or ion in the solution. From Poisson's equation (derived from Maxwell's first equation in Appendix C), we have

$$\nabla^2 \phi = -\rho(r)/\varepsilon = \frac{e}{\varepsilon} \sum_i z_i c_i(\infty) e^{-e z_i \phi(r)/kT}. \qquad (B.5)$$

Because of the spherical symmetry of the charged particle or ion within the electrolyte, the partial derivatives with respect to the spherical coordinates θ and ϕ are equal to zero, the partial derivative with respect to r can be written as a total derivative, and the Laplacian can be written as

$$\nabla^2 \phi = \frac{1}{r^2} \left[\frac{d}{dr} \left(r^2 \frac{d\phi}{dr} \right) \right]. \qquad (B.6)$$

It is a simple matter to show that the right hand side of this equation can be written as $(1/r)[d^2(r\,\phi(r))/dr^2]$, and Poisson's equation becomes

$$\frac{1}{r} \left[\frac{d^2(r\,\phi(r))}{dr^2} \right] = -\rho(r)/\varepsilon. \qquad (B.7)$$

Therefore, we have

$$\left[\frac{d^2(r\,\phi(r))}{dr^2} \right] = -\frac{e\,r}{\varepsilon} \sum_i z_i c_i(\infty) e^{-e z_i \phi(r)/kT}. \qquad (B.8)$$

This differential equation cannot be solved for the general case; however, in the event that the potential $\phi(r)$ is much less than

kT/e, $e\, z_i\, \phi(r)/kT \ll 1$. In this case, we can use the approximation $e^x = 1 + x$, where $x = e\, z_i\, \phi(r)/kT$. In this linear approximation, first introduced by Debye, we obtain

$$\frac{d^2(r\phi(r))}{dr^2} = -\frac{e\,r}{\varepsilon} \sum_i z_i\, c_i(\infty)\, (1 - e z_i\, \phi(r)/kT) \,. \qquad (B.9)$$

This reduces to

$$\frac{d^2(r\phi(r))}{dr^2} = \frac{e\,r}{\varepsilon} \sum_i z_i\, c_i(\infty)\, [e\, z_i\, \phi(r)/kT)] \,, \qquad (B.10)$$

since the first term on the right hand side of (B.9),

$$-(e\,r/\varepsilon) \sum_i z_i\, c_i(\infty) = 0,$$

because of overall electrical neutrality ($\sum_i z_i\, c_i(\infty) = 0$). To solve this differential equation, we let $u(r) = r\,\phi(r)$, and recognize that the right hand side is equal to a constant multiplied by $r\,\phi(r)$; that is, a constant multiplied by $u(r)$.

The constant, $(e^2/\varepsilon kT)\sum_i c_i(\infty)\, z_i^2$, has the dimensions of a reciprocal length (known as the *Debye length D*) squared. The resulting differential equation,

$$\frac{d^2 u(r)}{dr^2} = \frac{u(r)}{D^2}, \qquad (B.11)$$

has the solution

$$u(r) = A e^{-r/D} + B e^{+r/D} \,. \qquad (B.12)$$

With $u(r) = r\,\phi(r)$, we have

$$\phi(r) = \frac{1}{r} \left(A e^{-r/D} + B e^{+r/D} \right) , \qquad (B.13)$$

where

$$\frac{1}{D^2} = \frac{e^2}{\varepsilon kT} \sum_i c_i(\infty)\, z_i^2 \,. \qquad (B.14)$$

89

The boundary conditions are:
(1) $\phi(r) \to 0$, as $r \to \infty$, and
(2) $E_r = -\partial\phi(r)/\partial r = -d\phi(r)/dr = \sigma/\varepsilon$ at $r = a$ (see Appendix C).

The first boundary condition requires that $B = 0$, since $\phi(r)$ cannot become large without limit as $r \to \infty$. Now,

$$E_r = -\frac{\partial\phi(r)}{\partial r} = -\frac{d\phi(r)}{dr} = \frac{A}{r}\left[\left(\frac{1}{D} + \frac{1}{r}\right)e^{-r/D}\right]. \qquad \text{(B.15)}$$

Evaluating this expression at $r = a$, and solving for A, we obtain

$$A = \frac{Q}{4\pi\varepsilon}\left(\frac{e^{a/D}}{1 + a/D}\right),$$

since $E_{r=a} = [\sigma/\varepsilon]_{r=a} = Q/4\pi\,\varepsilon a^2$, where a is the radius of the charged particle or ion, and $\sigma = Q/4\pi a^2$.

Substituting these results, and recalling that $B = 0$, (B.13) becomes

$$\phi(r) = \frac{Qe^{-(r-a)/D}}{4\pi\varepsilon r(1 + a/D)}. \qquad \text{(B.16)}$$

The r component of the negative gradient of this expression gives the electric field. The spherically symmetric charge density $\rho(r)$ in the ionic cloud surrounding a charged particle or ion can also be obtained from the above results. From Poisson's equation as expressed in (B.7), we may write

$$\rho(r) = -\varepsilon\left[\frac{1}{r}\left(\frac{d^2(r\,\phi(r))}{dr^2}\right)\right]. \qquad \text{(B.17)}$$

From (B.11), with $u(r) = r\,\phi(r)$

$$\frac{d^2(r\phi(r))}{dr^2} = \frac{r\,\phi(r)}{D^2}, \qquad \text{(B.18)}$$

one obtains

$$\rho(r) = \frac{-\varepsilon}{r}\left(\frac{r\phi(r)}{D^2}\right) = \frac{-\varepsilon}{D^2}\phi(r). \qquad \text{(B.19)}$$

90

Finally,

$$\rho(r) = \frac{-Qe^{-(r-a)/D}}{4\pi D^2 r(1 + a/D)} .$$

(B.20)

This expression is needed to calculate the self-energy of Debye clouds in Chapter 2.

Returning to the expression for the Debye length D given in (B.14), we note that the dimensions of $c_i(\infty)$ are particles per unit volume, and we will want to express this in moles (mol) per liter. Dimensionally, moles per unit volume is equal to the number of particles per unit volume N/V divided by N_0 the number of particles per mole. Thus, if C_i is given in mol/V and N_0 is Avagadro's number, we have $c_i(\infty)\,(N/V) = C_i(\infty)\,(mol/V)\,N_0$.

It is useful and customary to define the *ionic strength* of an electrolyte according to the expression

$$I = \frac{1}{2}\sum_i C_i(\infty)\,z_i^2,$$

and with these changes, (B.14) becomes

$$\frac{1}{D^2} = \frac{2\,e^2}{\varepsilon kT}\,N_0\,I$$

(B.21)

Since all the constants are known, this equation may be used to obtain the Debye length as a function of ionic strength at temperature T.

Appendix C

Electrostatic stress

Maxwell's first equation relating the electric displacement D to the free charge ρ is $\nabla \cdot \mathbf{D} = \rho$. Since $\mathbf{D} = \varepsilon \mathbf{E}$, and $\mathbf{E} = -\nabla \phi$, we have $\nabla \cdot \mathbf{E} = -\nabla^2 \phi = \rho/\varepsilon$, and $\rho = -\varepsilon \nabla^2 \phi$ (Poisson's equation). Substituting this into (A.5), we obtain

$$U = -\varepsilon/2 \int \phi \nabla^2 \phi \, d\tau. \qquad (C.1)$$

We may recast this equation using the vector identity $\nabla \cdot (f \mathbf{A}) = f(\nabla \cdot \mathbf{A}) + \mathbf{A} \cdot \nabla f$, where f and $\mathbf{A}$ are scalar and vector functions respectively. If we let $f = \phi$, and $\mathbf{A} = \nabla \phi$, then $\phi \nabla^2 \phi = \nabla \cdot (\phi \nabla \phi) - (\nabla \phi)^2$, and

$$U = -\frac{\varepsilon}{2} \left[\int \nabla \cdot (\phi \nabla \phi) \, d\tau - \int (\nabla \phi)^2 \, d\tau \right]. \qquad (C.2)$$

Using the divergence theorem, $\int \mathbf{A} \cdot d\mathbf{A} = \int \nabla \cdot \mathbf{A} \, d\tau$, this may be written in terms of a surface integral over the surface bounding the volume τ

$$U = -\frac{\varepsilon}{2} \left[\int (\phi \nabla \phi) \cdot d\mathbf{A} - \int (\nabla \phi)^2 \, d\tau \right]. \qquad (C.3)$$

Since τ can be any volume that includes all the charge in the system, we may choose the bounding surface S to be at a great distance from the charge distribution. In the first integral ϕ

falls off as $1/r$, and therefore $\nabla\phi$ falls off as $1/r^2$, giving a total decay rate of $1/r^3$ for the integrand. Given that the surface area increases as r^2, the first integral decreases with distance as $1/r$ and can be made arbitrarily small by choosing a surface S sufficiently distant. Now $\mathbf{E} = -\nabla\phi$, and we are left with

$$U = \frac{\varepsilon}{2} \int E^2 \, d\tau. \qquad (C.4)$$

Note that the integrand, $\varepsilon E^2/2$, is the energy density in Joules per cubic meter (J/m^3) since $d\tau$ is a volume element. Recalling that 1 Joule equals 1 Newton-meter, we see that J/m^3 is dimensionally equivalent to N/m^2 (Newtons per square meter), and is also the electrostatic stress encountered in Chapter 2. From elementary electrostatics, the magnitude of the electric field E of a point charge Q (or a spherically symmetric charge distribution of radius R, charge Q, and $r \geq R$) is given by $Q/4\pi\varepsilon r^2 = Q/4\pi\varepsilon R^2$, for $r = R$, the radius of the sphere. Since the uniform charge density at the surface of the sphere σ is equal to $Q/4\pi R^2$, we see that $E(= E_r) = \sigma/\varepsilon$ at the surface of a uniformly charged sphere, and therefore

$$\varepsilon E^2/2 = \sigma^2/2\varepsilon. \qquad (C.5)$$

Appendix D

Equation 2.1

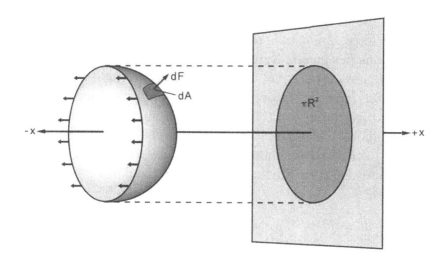

Fig. D.1. Mechanical equilibrium of a nuclear envelope. The model spherical half nuclear envelope is in equilibrium under the influence of surface tension acting to the left and electrostatic stress plus static pressure difference components acting to the right.

The hemispherical shell, representing half of the double membrane of the nuclear envelope (Figure D.1), is in equilibrium under the action of a uniform surface tension acting in the $-x$ direction from the rest of the membrane, and surface forces acting perpendicularly outward from the surface everywhere over the hemisphere. The surface forces arise from the pressure difference across the membrane and the surface electrical force per unit area of the membrane (membrane electrostatic stress).

The differential electrostatic force dF_e acting perpendicularly to an element of area dA of a charged surface can be expressed by the following equation [Griffiths, 1999a, Appendix C].

$$dF_e = (\varepsilon E^2/2)dA = (\sigma^2/2\,\varepsilon)dA, \qquad (D.1)$$

where E^2 is the magnitude squared of the electric field at dA, ε is the local permittivity just outside the surface at dA, and σ is the net surface charge per unit area. The model half nuclear envelope will be in equilibrium if the surface tension element of force, $2\gamma\,dl$ (where γ is the membrane surface tension in Newtons per meter, N/m) summed around the circumference of the membranes is equal to the total force in the $+x$ direction. The latter force is obtained by integrating the component in the $+x$ direction of the total differential force dF, which arises from membrane electrostatic stress plus the pressure difference $(p_1 - p_2)$ across the membranes. The factor of 2 arises because, as mentioned earlier, the nuclear envelope is a double membrane. Therefore,

$$2\gamma \int dl = \int [(p_1 - p_2)\, dA \cos\theta + (\sigma^2/2\,\varepsilon)\, dA \cos\theta], \qquad (D.2)$$

where θ is the angle between dF and the $+x$ axis. Since γ, $p_1 - p_2$, and $\sigma^2/2\,\varepsilon$ will be assumed constant over the membrane, we have

$$4\pi R\gamma = [(p_1 - p_2) + \sigma^2/2\varepsilon] \int dA \cos\theta. \qquad (D.3)$$

The quantity $dA \cos\theta$ is the projected area element on the yz plane, and the sum of such elements is the area of the projected circle, πR^2. Thus,

$$4\gamma/R = \Delta p + \sigma^2/2\,\varepsilon, \qquad (D.4)$$

where $\Delta p = (p_1 - p_2)$. This equation governs the equilibrium of the model half-nuclear envelope.

Appendix E

Equation 2.10

From Section 2.4, a necessary condition for the spontaneous splitting of a parent nucleus of radius R and charge Q into two daughter nuclei of radius R_1 and charge Q_D is $U_0 > U_1$. Therefore, we have

$$\frac{Q^2}{8\pi\varepsilon R} + 8\pi\gamma R^2 + \frac{1}{2}\int\limits_{R}^{\infty} \rho\,\phi\,d\tau >$$

$$2\left[\frac{Q_D^2}{8\pi\varepsilon R_1} + 8\pi\gamma R_1^2 + \frac{1}{2}\int\limits_{R_1}^{\infty} \rho\,\phi\,d\tau\right]. \qquad (\text{E.1})$$

Defining

$$U_{0a} = \frac{1}{2}\int\limits_{R}^{\infty} \rho\,\phi\,d\tau, \quad U_{1a} = 2\left[\frac{1}{2}\int\limits_{R_1}^{\infty} \rho\,\phi\,d\tau\right], \qquad (\text{E.2})$$

and $\Delta U_a = U_{0a} - U_{1a}$, we may write

$$\Delta U_a + \frac{Q^2}{8\pi\varepsilon R} + 8\pi\gamma R^2 > 2\left[\frac{Q_D^2}{8\pi\varepsilon R_1} + 8\pi\gamma R_1^2\right]. \qquad (\text{E.3})$$

To perform the integrals in the ΔU_a term, we note the following relations from Debye theory [Benedek and Villars, 2000b;

Appendix B]:

$$\phi(r) = \frac{Qe^{-(r-R)/D}}{4\pi\varepsilon r(1+R/D)}, \text{ and } \rho(r) = \frac{-Qe^{-(r-R)/D}}{4\pi D^2 r(1+R/D)},$$

where r is the distance from the center of the nucleus of radius R. Suppressing the minus sign to obtain the magnitude, we have

$$U_{0a} = \frac{1}{2} \int_R^\infty \frac{Q^2\,e^{-2(r-R)/D}}{16\,\pi^2\,\varepsilon\,D^2\,(1+R/D)^2\,r^2}\left(4\,\pi\,r^2\right)d\,r, \qquad \text{(E.4)}$$

where the spherically symmetric volume element $d\tau = 4\,\pi\,r^2\,dr$. Noting that the r^2 terms cancel, and after further simplification, we have

$$U_{0a} = \frac{Q^2\,e^{2R/D}}{8\,\pi\,\varepsilon\,D^2(1+R/D)^2} \int_R^\infty e^{-2r/D}\,d\,r. \qquad \text{(E.5)}$$

Evaluating this integral, we obtain

$$U_{0a} = \frac{D\,Q^2}{16\,\pi\,\varepsilon\,D^2(1+R/D)^2} = \frac{D\,Q^2}{16\,\pi\,\varepsilon\,R^2}, \qquad \text{(E.6)}$$

since $D \ll R$, and $1+R/D \to R/D$. Writing this as

$$U_{0a} = \frac{D}{2\,R}\left(\frac{Q^2}{8\,\pi\,\varepsilon R}\right), \qquad \text{(E.7)}$$

and recalling that $D = 1$ nm, $R = 5$ μm, and $D/R \ll 1$, we see that the energy of the ionic cloud associated with the parent nucleus is completely negligible compared to the self-energy, $Q^2/8\pi\varepsilon R$.

An entirely similar calculation for U_{1a} yields

$$U_{1a} = \frac{D}{R_1}\left(\frac{Q_D^2}{8\,\pi\,\varepsilon R_1}\right). \qquad \text{(E.8)}$$

Therefore, since $\Delta U_a = U_{0a} - U_{1a}$, we may drop this term from (E.3). Replacing R_1 by R/β and Q_D by Q/β^2 (since Q_D is proportional to R_1^2), where β is the ratio of R to R_1, we may write (E.3) as

$$\frac{Q^2}{8\pi\varepsilon R}\left(1 - \frac{2}{\beta^3}\right) > 8\,\pi\,\gamma\,R^2\left(\frac{2}{\beta^2} - 1\right). \qquad \text{(E.9)}$$

By definition, $Q = 4\pi\sigma R^2$, so this may be written

$$\frac{\sigma^2}{2\varepsilon}\left(1 - \frac{2}{\beta^3}\right) > \frac{2\,\gamma}{R}\left(\frac{2}{\beta^2} - 1\right). \qquad \text{(E.10)}$$

Appendix F

Bibliography

Aist J.R., Bayles C.J., Tao W., Berns M.W. 1991. Direct experimental evidence for the existence, structural basis and function of astral forces during anaphase B *in vivo*. J. Cell Sci. **100**:279.

Alberts B., Bray D., Lewis J., Raff M., Roberts M.K., Watson J.D. 1994a. Molecular Biology of the Cell. New York: Garland Publishing Company, p. 920.

Alberts B., Bray D., Lewis J., Raff M., Roberts M.K., Watson J.D. 1994b. Molecular Biology of the Cell. New York: Garland Publishing Company, p. 1041.

Alberts B., Bray D., Lewis J., Raff M., Roberts M.K., Watson J.D. 1994c. Molecular Biology of the Cell. New York: Garland Publishing Company, p. 930.

Alberts B., Bray D., Lewis J., Raff M., Roberts M.K., Watson J.D. 1994d. Molecular Biology of the Cell. New York: Garland Publishing Company, p. 926.

Alexander S.P., Rieder C.L. 1991. Chromosome motion during attachment to the vertebrate spindle: Initial saltatory-like behavior of chromosomes and quantitative analysis of force production by nascent kinetochore fibers. J. Cell Biol. **113**:805.

Amirand C. *et al.* 2000. Intracellular pH in one-cell mouse embryo differs between subcellular compartments and between interphase and mitosis. Biol. Cell **92**:409.

Anderson R.G.W., Jacobson K. 2002. Cell biology: A role for lipid shells in targeting proteins to caveolae, rafts, and other rafts, and other lipid domains. Science **296**:1821.

Baker N.A., Sept D., Joseph S., Holst M.J., McCammon J.A. 2001. Electrostatics of nanosystems: Applications to microtubules and the ribosome. Proc. Nat. Acad. Sci. **98**:10037.

Beaudouin J. *et al.* 2002. Nuclear envelope breakdown proceeds by microtubule-inducing tearing of the lamina. Cell **108**:83.

Benedek G.B., Villars F.M.H. 2000a. Physics: With Illustrative Examples From Medicine and Biology: Electricity and Magnetism. New York: Springer-Verlag, p. 403.

Benedek G.B., Villars F.M.H. 2000b. Physics: With Illustrative Examples From Medicine and Biology: Electricity and Magnetism. New York: Springer-Verlag, p. 408.

Benedek G.B., Villars F.M.H. 2000c Physics: With Illustrative Examples From Medicine and Biology: Electricity and Magnetism. New York: Springer-Verlag, p. 400.

Bergen L.G., Kuriyama R., Borisy G.G. 1980. Polarity of microtubules nucleated by centrosomes and chromosomes of Chinese hamster ovary. J. Cell Biol. **84**:151.

Bockris J.O., Reddy A.K.N. 1977. Modern Electrochemistry. New York: Plenum Press.

Borisy G.G., Olmsted J.B. 1972. Nucleated assembly of microtubules in porcine brain extracts. Science **177**:1196.

Brown J.A., Tuszyński J.A. 1997. Dipole interactions in axonal microtubules as a mechanism of signal propagation. Phys. Rev. E **56**:5834.

Cande W.Z. 1981. In: H.G. Schweiger (Ed.), International Cell Biology. Berlin: Springer Publishing Company, pp. 382-391.

Civelekoglu-Scholey G., Sharp D.J., Mogilner A., Scholey J.M. 2006. Model of chromosome motility in Drosophila embryos: Adaptation of a general mechanism for rapid mitosis. Biophys. J. **90**:3966.

Cleveland L.R. 1953. Trans. Am. Phil. Soc. **43**:809.

Cooper M.S. 1979. Long-range dielectric aspects of the eukaryotic cell cycle. Physiol. Chem. Phys. **11**:435.

Cowley A.C., Fuller N.L., Rand R.P., Parsegian V.A. 1978. Measurement of repulsive forces between charged phospholipid bilayers. Biochemistry **17**:3163.

De Brabander M., Geuens G., Nuydens R. 1982. Microtubule stability and assembly in living cells: The influence of metabolic inhibitors, taxol and pH. Cold Spring Harbor Symp. Quant. Biol. **46**:227.

Deery W.J., Brinkley B.R. 1983. Cytoplasmic microtubule-disassembly from endogenous tubulin in a Brij-lysed cell model. J. Cell Biol. **96**:1631.

Diebler H., Eigen G., Ilgenfritz G., Maass G., Winkler R. 1969. Kinetics and mechanism of reactions of main group metal ions with biological carriers. Pure Appl. Chem. **20**:93.

Ellenberg J. *et al.* 1997. Nuclear membrane dynamics and reassembly in living cells: Targeting of an inner nuclear membrane protein in interphase and mitosis. J. Cell Biol. **138**:1193.

Euteneuer U., McIntosh J.R. 1981. Structural polarity of kinetochore microtubules in PtK1 cells. J. Cell Biol. **89**:338.

Fettiplace R., Andrews D.M., Haydon D.A. 1971. The thickness, composition and structure of some lipid bilayers and natural membranes. J. Membrane Biol. **5**:277.

Frankl D.R. 1986. Electromagnetic Theory. Englewood Cliffs, New Jersey: Prentice-Hall, p. 221.

Gagliardi L.J. 2002a. Electrostatic force in prometaphase, metaphase, and anaphase-A chromosome motions. Phys. Rev. E **66**:011901.

Gagliardi L.J. 2002b. Microscale electrostatics in mitosis. J. Electrostat. **54**:219.

Gagliardi L.J. 2005. Electrostatic force generation in chromosome motions during mitosis. J. Electrostat. **63**:309.

Gagliardi L.J. 2008. Induced electrostatic charge in poleward motion of chromosomes during mitosis. J. Electrostat. **66**:147.

Gerace L., Blobel G. 1980. The nuclear envelope lamina is reversibly depolymerized during mitosis. Cell **19**:277.

Giese A.C. 1968. Cell Physiology. Philadelphia: W. B. Saunders Publishing Company, p. 104.

Glick M.C., Gerner E.W., Warren L. 1971. Changes in the carbohydrate content of the KB cell during the growth cycle. J. Cell Physiol. **77**:1.

Gorbsky G.J., Sammak P.J., Borisy G.G. 1987. Chromosome move poleward in anaphase along stationary microtubules that coordinately disassemble from their kinetochore ends. J. Cell Biol. **104**:9.

Grancell A., Sorger P.K. 1998. Chromosome movement: Kinetochores motor along. Current Biol. **8**:R382.

Griffiths D.J. 1999a. Introduction to Electrodynamics. Upper Saddle River, New Jersey: Prentice-Hall, p. 103.

Griffiths D.J. 1999b. Introduction to Electrodynamics. Upper Saddle River, New Jersey: Prentice-Hall, p. 75.

Grishchuk E.L. *et al.* 2008a. The Dam1 ring binds microtubules strongly enough to be a processive as well as energy-efficient coupler for chromosome motion. Proc. Natl. Acad. Sci. USA **105**:15423.

Grishchuk E.L. *et al.* 2008b. Different assemblies of the DAM1 complex follow shortening microtubules by distinct mechanisms. Proc. Natl. Acad. Sci. USA **105**:6918.

Guimaraes G.J., Dong Y., McEwen B.F., DeLuca J.G. 2008. Kinetochore-microtubule attachment relies on the disordered N-terminal tail domain of Hec1. Current Biol. **18**:1778.

Hartley G.S., Roe J.W. 1940. Ionic concentrations at interfaces. Trans. Faraday Soc. **35**:101.

Hasted J.B., Ritson D.M., Collie C.H. 1948. Dielectric properties of aqueous ionic solutions. Parts I and II. J. Chem. Phys. **16**:1.

Hays T.S., Salmon E.D. 1990. Poleward force at the kinetochore in metaphase depends on the number of kinetochore microtubules. J. Cell Biol. **110**:391.

Heald R., Tournebize R., Blank T., Sandaltzopoulos R., Becker

P., Hyman A., Karsenti E. 1996. Self-organization of micro-tubules into bipolar spindles around artificial chromosomes in Xenopus egg extracts. Nature **382**:420.

Heinz W.F., Hoh J.H. 1999. Relative surface charge density mapping with the atomic force microscope. Biophys. J. **76**:528.

Hepler P.K., Callaham D.A. 1987. Free calcium increases in anaphase in stamen hair cells of Tradescantia. J. Cell Biol. **105**: 2137.

Hepler P.K. 1989. In: J.S. Hyams and B.R. Brinkley (Eds), Mitosis: Molecules and Mechanisms. San Diego: Academic Press, pp. 241-271.

Hormeño S. *et al.*. 2009. Single centrosome manipulation reveals its electric charge and associated dynamic structure. Biophys. J. **97**:1022.

Inoue S., Salmon E.D. 1995. Force generation by microtubule assembly/disassembly in mitosis and related movements. Mol. Biol. Cell **6**:1619.

Israelachvili J.N. 1991. Intermolecular and Surface Forces. London: Academic Press.

Jaffe L.F., Nuccitelli R. 1977. Electrical controls of development. Ann. Rev. Biophys. Bioeng. **6**:445.

Jordan-Lloyd D., Shore A. 1938. The Chemistry of Proteins, London: J. A. Churchill Publishing Company.

Joshi H.C., Palacios M.J., McNamara L., Cleveland D.W. 1992. γ-Tubulin is a centrosomal protein required for cell cycle-dependent microtubule nucleation. Nature **356**:80.

Kiehart D.P. 1981. Studies on the *in vivo* sensitivity of spindle microtubules to calcium ions and evidence for a vesicular calcium-sequestereing system. J. Cell Biol. **88**:604.

Kirschner M.W. 1980. Implications of treadmilling for the stability and polarity of actin and tubulin polymers in vivo. J. Cell Biol. **86**:330.

Lewin B.M. 1978. Gene Expression: Eucaryotic Chromosomes, vol. 2. New York: John Wiley and Sons, pp. 34-36.

Maddox P., Desai A., Oegema K., Mitchison T.J., Salmon E.D. 2002. Poleward microtubule flux is a major component of spindle dynamics and anaphase A in mitotic Drosophila embryos. Current Biology **12**:1670.

Maiato H., DeLuca J., Salmon E.D., Earnshaw W.C. 2004. The dynamic kinetochore-microtubule interface. J. Cell Science **117**: 5461.

Marshall I.C.B., Wilson K.L. 1997. Nuclear envelope assembly after mitosis. Trends Cell Biol. **7**:69.

Masuda H., McDonald K.L., Cande W.Z. 1988. The mechanism of anaphase spindle elongation: Uncoupling of tubulin incorporation and microtubule sliding during *in vitro* spindle reactivation. J. Cell Biol. **107**:623.

Mayhew E. 1966. Cellular electrophoretic mobility and the mitotic cycle. J. Gen. Physiol. **49**:717.

Mayhew E., O'Grady E.A. 1965. Electrophoretic mobilities of tissue culture cells in exponential and parasynchronous growth. Nature **207**:86.

McIntosh J.R. *et al.* 2007. Kinetochore-Microtubule Interactions Visualized by EM Tomagraphy, presented at the 47th Annual Meeting of the American Society for Cell Biology, Washington, DC, December 1-5, 2007.

Mitchison T.J., Evans L., Schulze E., Kirschner M. 1986. Sites of microtubule assembly and disassembly in the mitotic spindle. Cell **45** :515.

Mitchison T.J. 1989. Polewards microtubule flux in the mitotic spindle: Evidence from photoactivation of fluorescence. J. Cell Biol. **109**:637.

Mitchison T.J., Salmon E.D. 1992. Poleward kinetochore fiber movement occurs during both metaphase and anaphase-A in newt lung cell mitosis. J. Cell Biol. **119**:569.

Newport J., Spann T. 1987. Disassembly of the nucleus in mitotic extracts: Membrane vesicularization, lamin disassembly, and chromosome condensation are independent processes. Cell **48**:219.

Nicklas R.B. 1975. In: S. Inoue and R.E. Stephens (Eds.), Molecules and cell movement. New York: Raven Press, pp. 97-117.

Nicklas R.B. 1983. Measurements of the force produced by the mitotic spindle in anaphase. J. Cell Biol. **97**:542.

Nicklas R.B., Kubai D.F. 1985. Microtubules, chromosome movement, and reorientation after chromosomes are detached from the spindle by micromanipulation. Chromosoma **92**:313.

Nicklas R.B. 1987. In: J.P. Gustafson, R. Appels, and R.J. Kaufman (Eds.), Chromosome Structure and Function. New York: Plenum Publishing Company, pp. 53-74.

Nicklas R.B. 1988. The forces that move chromosomes in mitosis. Ann. Rev. Biophys. Biophys. Chem. **17**:431.

Nicklas R.B. 1989. The motor for poleward chromosome movement in anaphase is in or near the kinetochore. J. Cell Biol. **109**:2245.

Olmsted J.B., Borisy G.G. 1975. Ionic and nucleotide requirements for microtubule polymerization in vitro. Biochemistry **14**:2996.

Pashley R.M. 1981. Hydration forces between mica surfaces in aqueous electrolyte solutions. J. Colloid Interface Sci. **83**:153.

Pauling L. 1945. The adsorption of water by proteins. J. Am. Chem. Soc. **67**:555.

Pickett-Heaps J. 1991. Int. Rev. Cytol. **128**:63.

Poenie M., Alderton J., Tsien R.Y., Steinhardt R.A. 1985. Changes of free calcium levels with stages of the cell division cycle. Nature **315**:147.

Pollack G.H. 2001. Cells, Gels and the Engines of Life. Seattle: Ebner and Sons Publishers, p. 69.

Rieder C.L. 1982. The formation, structure, and composition of the mammaliam kinetochore and kinetochore fiber. Int. Rev. Cytol. **79**:1.

Rieder C.L, Davison E.A, Jensen L.C.W. 1986. Oscillatory movements of monooriented chromosomes and their position relative to the spindle pole result from the ejection properties of the aster and half-spindle. J. Cell Biol. **103**:581.

Rieder C.L., Alexander S.P. 1990. Kinetochores are transported poleward along a single astral microtubule during chromosomes attachment to the spindle in newt lung cells. J. Cell Biol. **110**:81.

Sackett D. 1997. pH-induced conformational changes in the carboxy-terminal tails of tubulin, presented at the Banff Workshop Molecular Biophysics of the Cytoskeleton, Banff, Alberta, Canada, August 25-30, 1997.

Saito M. *et al.* 2002. Age-dependent reduction in sialidase activity of nuclear membranes from mouse brain. Exp. Gerentol. **37**:937.

Salina D. *et al.* 2002. Cytoplasmic dynein as a facilitator of nuclear envelope breakdown. Cell **108**:97.

Salmon E.D. 1975. Spindle microtubules: thermodynamics of in vivo assembly and role in chromosome movement. Ann. N.Y. Acad. Sci.**253**:383.

Salmon E.D., Segall R.R. 1980. Calcium-labile mitotic spindles isolated from sea urchin eggs (Lytechinus variegatus). J. Cell Biol. **86**:355.

Salmon E.D. 1989. In: J.S. Hyams and B.R. Brinkley (Eds.), Mitosis: Molecules and Mechanisms. San Diego: Academic Press, pp. 119-181.

Satarić M.V., Tuszyński J.A., Žakula R.B. 1993. Kinklike excitations as an energy-transfer mechanism in microtubules. Phys. Rev. E **48**:589.

Schatten G., Bestor T., Balczon R. 1985. Intracellular pH shift leads to microtubule assembly and microtubule-mediated motility during sea urchin fertilization: Correlations between elevated intracellular pH, microtubule activity and depressed intracellular pH and microtubule disassembly. Eur. J. Cell Biol. **36**:116.

Schelkunoff S.A. 1963. Electromagnetic Fields. New York: Blaisdell Publishing Company, p. 29.

Seaman G.V.F., Cook G.M.W. 1965. In: E.J. Ambrose (Ed.), Cell Electrophoresis. London: J. A. Churchill Publishing Company, pp. 49-65.

Segal J.R. 1968. Surface charge of giant axons of squid and lobster. Biophys. J. **8**:470.

Silver R.B. 1989. Nuclear envelope breakdown and mitosis in sand dollar embryos is inhibited by microinjection of calcium buffers in a calcium-reversible fashion, and by antagonists of intracellular Ca^{2+} channels. Dev. Biol. **131**:11.

Silver R.B. 1994. Time-resolved imaging of Ca^{2+}-dependent aequorin luminescence of microdomains and QEDs in synaptic preterminals. Biol. Bull. **187**:293.

Silver R.B., Reeves A.P., Kelley B.P., Fripp W.J. 1996. Inherent nonrandom structure of the pre-nuclear envelope breakdown calcium signal in sand dollar (Echinaracnius parma) embryos. Biol. Bull. **191**; 278.

Silver R.B., King L.A., Wise A.F. 1998. Calcium regulatory endomembranes of the prophase mitotic apparatus of sand dollar cells contain enzyme activities that produce leukotriene B4 but not 1,4,5-inositol trisphosphate. Biol. Bull. **195**:209.

Steinhardt R.A., Morisawa M. 1982. In: Intracellular pH: Its Measurement, Regulation, and Utilization in Cellular Functions, R. Nuccitelli and D.W. Deamer (Eds.). New York: Alan R. Liss, pp. 361-374.

Stick R., Schwartz H. 1983. Disappearance and reformation of the nuclear lamina structure during specific stages of meiosis in oocytes. Cell **33**:949.

Stracke R., Böhm K.J., Wollweber L., Tuszyński J.A., Unger E. 2002. Analysis of the migration behaviour of single microtubules in electric fields. Bioch. and Biophys. Res. Comm. **293**:602.

Stricker SA. 1995. Time-lapse confocal imaging of calcium dynamics in starfish embryos. Dev. Biol. **170**:496.

Tamura A, Morita K, Fujii T, Kojima K. 1982. Cell Struct. and Funct. **7**:21.

Terasaki M. *et al.* 2001. A new model for nuclear envelope breakdown. Mol. Biol. Cell **12**:503.

Teschke O., Ceotto G., de Souza E.F. 2001. Interfacial water dielectric-permittivity-profile measurements using atomic force microscopy. Phys. Rev. E **64**:011605.

Tippit D.H., Schultz D., Pickett-Heaps J.D. 1978. Analysis of the distribution of spindle microtubules in the diatom Fragilaria J. Cell Biol. **79**:737

Toney M.F., Howard J.N., Richer J., Borges G.L., Gordon J.G, Melroy OR, Wiesler DG, Yee D, Sorensen L. 1994. Voltage-dependent ordering of water molecules at an electrode-electrolyte interface. Nature **368**:444.

Tuszyński J.A., Hameroff S., Satarić M.V., Trpisová B., Nip M.L.A. 1995. Ferroelectric behavior in microtubule dipole lattices: Implications for information processing, signaling and assembly/disassembly. J. Theor. Biol. **174**:371.

Tuszyński J.A., Brown J.A., Hawrylak P. 1998. Dielectric polarization, electrical conduction, information processing and quantum computation in microtubules. Are they plausible? Phil. Trans. R. Soc. Lond. **A356**:1897.

Warren L., Zeidman I., Buck C.A. 1975. The surface glycoproteins of a mouse melanoma growing in culture and as a solid tumor in vivo. Cancer Res. **35**:2186.

Weisenberg R.C. 1972. Microtubule formation *in vitro* in solutions containing low calcium concentrations. Science **177**:1104.

Weiss L. 1968. Studies on cell deformability. V. Some effects of ribonuclease. J. Theoret. Biol. **18**:9.

Weiss L. 1969. In: G.H. Bourne and J.F. Danielli (Eds.), Int. Rev. Cytol. **26**. New York: Academic Press, pp. 63-105.

Westermann S. *et al.*. 2005. Formation of a dynamic kinetochore-microtubule interface through assembly of the Dam1 ring complex. Molecular Cell **17**:277.

Wolfe S.L. 1993a. Molecular and Cellular Biology. Belmont, CA: Wadsworth Publishing Company, p. 1012.

Wolfe S.L. 1993b. Molecular and Cellular Biology. Belmont, CA: Wadsworth Publishing company, p. 422.

Wolfe S.L. 1993c. Molecular and Cellular Biology. Belmont, CA: Wadsworth Publishing Company, p. 425.

Wolfe S.L. 1993d. Molecular and Cellular Biology. Belmont, CA: Wadsworth Publishing Company, p. 1028.

Zhang D.H., Callaham D.A., Hepler P.K. 1990. Regulation of anaphase chromosome motion in Tradescantia stamen hair cells by calcium and related signalling agents. J. Cell Biol. **111**:171.

Zhang D.H., Chen W. 2003. Dynamics of severed kinetochore fiber stubs in the cytoplasm of anaphase grasshopper spermatocytes, presented at the 43rd Annual Meeting of the American Society for Cell Biology, San Francisco, December 13-17, 2003.

Index

116

Printed in the United States
by Baker & Taylor Publisher Services